Beyond Fear:
The Truth About Nuclear Energy

History, Science, And Geopolitics Of
The World's Most Feared And Misunderstood Energy

Why The World Needs To Rethink Nuclear
Before It's Too Late

By João Garcia Pulido

Who Should Read This Book

This book is for everyone.

For the curious.

For the skeptic.

For the one who always believed — and for the one who never wanted to hear about it.

It's for those who seek to understand — not to repeat.

For those who value science over noise.

For those who still believe that quality information can be a bridge — not a trench.

This book is not neutral in thought — it is neutral in manipulation.

It is not impartial in the pursuit of truth — but it is free from agendas that distort and oversimplify.

There is no ideology served here — only knowledge.

And like all true knowledge, it unsettles, provokes, and questions.

It is a book for all who have an open mind and the humility to hear what has never been clearly told.

It is also an invitation — almost a challenge — to those who grew up immersed in rigid narratives, inherited prejudices, and constant media intoxication.

Perhaps this book will not change convictions — but if it can plant a fertile doubt, it will have done its job.

Agreement is not required.

But respect is demanded — for the work presented here is the result of serious research, critical thinking, and a commitment to truth, however uncomfortable.

If you're willing to think for yourself — then this book is yours.

Acknowledgments

Writing this book has been one of the most intense and transformative journeys of my life. But no great journey is made alone.

First and foremost, I thank my family — for the time I stole from them, for the hours I was present but absent, immersed in thoughts, pages, and revisions. Without your patience and silent love, this book would never have come to life.

To my colleagues and longtime friends who, over decades in the energy sector, encouraged, challenged, and inspired me — your confidence gave me strength whenever fatigue tried to win.

To Hillshire Media, my publisher, for having the courage to believe in this project from the very beginning and for giving me the freedom to write with truth, precision, and passion.

To you, dear reader, a heartfelt thank you — for by purchasing this book, you give me encouragement and purpose. More than a commercial gesture, your choice is an act of trust. I hope to live up to it.

And finally, a heartfelt and brotherly thanks to my colleague and companion on this journey, Bruno Cerqueira. Your presence was constant — in meticulous revisions, in the creation of the diagrams and images that enrich these pages, and in the unwavering support during long nights and exhausting days. This book is yours, too.

This book is born from a commitment to truth to science, and to a more sustainable future. May its pages inspire the change our world so urgently needs.

Table of Contents

Introduction –
Nuclear Energy:
Understanding Before Deciding

We live in an era of urgent decisions. The climate crisis is advancing. Fossil fuels, although still dominant, are becoming increasingly unsustainable. The promise of renewables is real, but their intermittency and logistical complexity limit it. Electricity has become the lifeblood of modern civilization — and the question that looms over governments, companies, and citizens alike is: **where will the energy that powers the future come from?**

This book was written to offer an honest, well-documented, and accessible answer to that question — an answer often ignored due to misinformation, ideology, or fear: **nuclear energy**.

More than technology, nuclear power represents a frontier of human knowledge. It is pure science applied to collective well-being. It is precision engineering in the service of energy stability. It is a tool of power and, at the same time, an opportunity for peace. Yet, it is also a subject burdened with controversy, myths, and mistrust. That is why this book is not only technical. It is also historical, political, strategic, and deeply human.

Throughout the following pages, the reader will be taken on a journey that begins with an ancestral reflection on the relationship between humankind and energy, then moves through the discovery of the atom, the evolution of nuclear physics, and the defining moments of the 20th century — from the Manhattan Project to the launch of the "Atoms for Peace" program.

We will explore nuclear accidents in depth — neither hiding nor exaggerating them. We will analyze radioactive waste with data and real solutions. We will dismantle anti-nuclear arguments based on facts, not ideologies. We will also present the peaceful applications of nuclear technology — from medicine to agriculture, from desalination to hydrogen production.

This book also delves into the geopolitics of energy, illustrating how access to and mastery of nuclear technology influence the world's outstanding power balances. We will see how the nations that invested in atomic energy have thrived — and how those that abandoned it now face energy crises and external dependency.

In the final section, we explore the role of nuclear energy in the energy transition, with a focus on new technologies such as Small Modular Reactors (SMRs), breakthroughs in nuclear fusion, the use of artificial intelligence and new materials, and the growing link between energy, strategic minerals, and national sovereignty.

The reader will find here not just a collection of facts but a vision for the future. A clear proposition: nuclear energy is essential to ensuring a sustainable planet, a stable economy, and a society free from energy extortion.

This book was designed for a broad audience. It does not require a technical background, but it offers scientific rigor. It does not impose a truth but proposes a serious debate. It is an invitation to reflect — and, for many, perhaps a necessary provocation. After all, **ignorance is comfortable, but knowledge is liberating**.

If the reader is willing to question, to hear the other side, and to consider that nuclear power is not only possible — but necessary — then this book will find its place. Welcome to the discussion that may shape the 21st century.

Chapter 1 –
Energy and Humanity:
From Firewood to Nuclear

Since the dawn of civilization, humanity has always depended on energy to survive and evolve.

Fire, discovered by our ancestors hundreds of thousands of years ago, marked the first significant milestone in the use of energy for purposes beyond mere survival. The heat generated by fire not only allowed food to be cooked but also provided warmth in cold regions, protecting entire populations from harsh climates. Fire also played a crucial role in nighttime illumination, allowing human activity to extend beyond the limits of daylight. Moreover, it offered protection against fierce predators, ensuring greater safety for early human settlements.

As societies evolved, the use of energy became increasingly sophisticated. The agricultural revolution, approximately 10,000 years ago, marked a significant turning point in human history, enabling populations to transition from a nomadic to a sedentary lifestyle. Solar energy was harnessed to grow food, while domesticated animals provided muscular power to plough the land, significantly boosting agricultural productivity.

With the rise of the first organized civilizations, the energy demand grew exponentially. The invention of the wheel and the development of metallurgy were fundamental advances that required new energy sources, such as charcoal and wind-powered furnaces. During the Classical Age, the Greeks and

Romans already made use of hydropower to operate mills and improve industrial production — from flour to textiles.

The Industrial Revolution in the 18th century marked the beginning of a new era of energy. Coal came to be used on a large scale to fuel steam engines, revolutionizing transport, and industrial manufacturing. The 20th century saw the discovery of oil and electricity as primary energy sources, facilitating the expansion of cities, the evolution of transportation systems, and the development of new technologies.

However, the exponential growth in energy demand also brought challenges. The burning of fossil fuels has caused significant environmental impacts, including global warming and air pollution. In light of this, the search for cleaner and more efficient energy sources has become an international priority.

In this context, nuclear energy emerges as one of the most promising alternatives. With its capacity for clean and efficient generation — and no direct greenhouse gas emissions — it presents itself as a viable solution to meet humanity's growing energy needs.

In the chapters ahead, we will explore the history of nuclear energy, its peaceful applications, challenges, and global impacts, analyzing how this source could shape the future of civilization.

The First Sources of Energy

Wood was the first significant source of energy used by humanity. For thousands of years, it has served as the primary fuel for cooking, generating heat, and even driving rudimentary industrial processes. Its abundance and ease of access made it the dominant option until new sources were discovered.

With the advancement of metallurgy and growing energy demand, wood began to be replaced by charcoal, which provided a more efficient and higher-temperature combustion process. Charcoal is produced through the controlled burning of wood in a process known as pyrolysis, where wood is heated in a low-oxygen environment, resulting in a fuel with greater calorific value. This development was essential for metalworking, allowing the production of stronger and more advanced tools.

Mineral coal, on the other hand — formed over millions of years by the decomposition of organic matter buried under high pressure and temperature — began to be widely used from the Industrial Revolution onwards. Unlike charcoal, which depends on forest exploitation, mineral coal is extracted from underground or open-pit mines. Its superior energy content and large-scale availability enabled the creation of steam engines, locomotives, and a more robust industrial sector. This profoundly transformed society, accelerating urbanization and economic growth but also giving rise to an energy exploitation model with severe environmental consequences.

The Industrial Revolution and the Rise of Coal as the Main Fuel

In the 18th century, the Industrial Revolution transformed the way energy was used. Mineral coal became the dominant energy source, powering the steam engines that drove factories, transport systems, and electricity generation. Cities grew rapidly, and the demand for coal soared, leading to intensive mining operations.

Although coal was the driving force behind industrialization, it also brought environmental challenges. Burning coal releases large amounts of carbon dioxide (CO_2) and atmospheric pollutants, contributing to air pollution and the intensification of the greenhouse effect. Moreover, working conditions in the mines were hazardous, with frequent collapses and exposure to toxic substances.

The Discovery and Expansion of Oil as the Dominant Energy Source in the 20th Century

At the end of the 19th century and the beginning of the 20th, humanity witnessed its first major modern energy transition: the shift from coal to oil as the dominant energy source. This change brought political, economic, and strategic challenges and, much like today, sparked heated debates between supporters and opponents.

For over 70 years, coal has been the primary energy source for both the industrial and military sectors. However, as energy efficiency and consumption intensity increased, it became clear that a more flexible and efficient source of energy was needed. Oil emerged as that alternative, offering significant advantages over coal, including higher energy density, easier storage, and reduced labor requirements for handling.

In the early 20th century, the Royal Navy depended almost exclusively on coal for its energy needs. Coal was abundant in the United Kingdom, and the mining and transportation infrastructure was well established. However, coal has had operational disadvantages, such as requiring large crews to load it, difficulty in refueling at sea, and producing smoke that could reveal the position of ships.

Oil began to emerge as a superior alternative. Oil-powered ships were faster, had greater range, and required less labor for operation and maintenance. Moreover, oil allowed for refueling at sea — a critical advantage in military operations.

The United Kingdom, then a leading industrial and military power, became the epicenter of this transition. The modernization of the British fleet was led by Winston Churchill, who served as First Lord of the Admiralty from 1911 to 1915. Churchill recognized that oil-powered ships had a decisive operational advantage: they were faster, more efficient, and required less maintenance than coal-powered ones.

Churchill encountered considerable opposition within both the government and the Royal Navy, which were concerned about relying on foreign oil, as the UK lacked substantial domestic reserves. To address this risk, he brokered a strategic deal with the Anglo-Persian Oil Company (now BP), ensuring direct access to Middle Eastern oil. This move not only modernized the British navy but also played a pivotal role in shaping the geopolitics of oil throughout the 20th century.

The transition to oil solidified the Royal Navy's position as the world's most powerful naval force during World War I. Moreover, the agreement with Anglo-Persian laid the foundation for the United Kingdom's strong presence in the Middle East, influencing global energy policy to this day.

Churchill's intervention in shifting the Royal Navy from coal to oil was a visionary decision that blended technological innovation, military strategy, and economic diplomacy. His ability to foresee the advantages of oil and overcome internal resistance was crucial to the transition's success, consolidating British naval power and shaping the course of global energy history.

This shift also sets a precedent for future energy transitions. Just as coal was replaced by oil, today we debate the replacement of fossil fuels with cleaner and more sustainable sources. The lesson from history, however, is that every successful energy transition must consider factors such as energy security, economic costs, and technological efficiency.

By the turn of the 20th century, oil had emerged as a new and powerful source of energy. The invention of the internal combustion engine and the growing demand for more efficient fuels led to its widespread adoption. Oil quickly became the backbone of the global economy, powering cars, airplanes, ships, and electricity production.

Oil exploration and refining enabled the production of a wide range of products — from fuels like gasoline and diesel to plastics and chemicals. This energy resource transformed the global economy and geopolitics, granting oil-producing countries significant influence and economic power.

However, oil also brought its challenges. Like coal, its combustion generates greenhouse gas emissions, contributing to climate change.

Chart 1. Evolution of CO_2 Emissions

Growth of CO_2 Emissions in the 20th and 21st Centuries

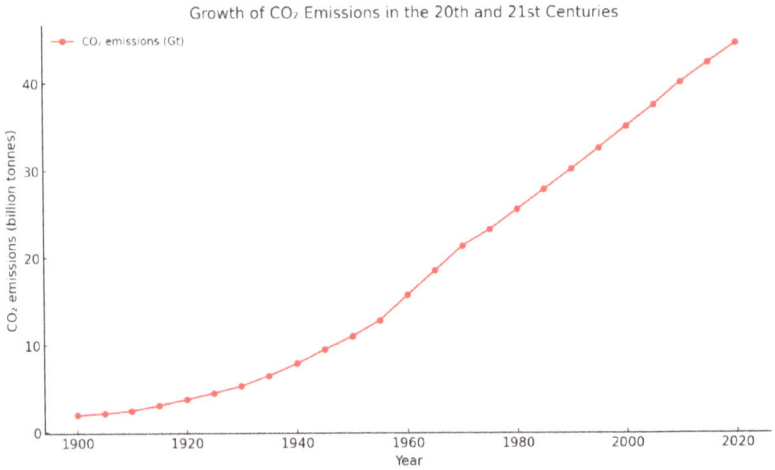

Source: Own elaboration based on the data presented in the Summary
Table at the end of this chapter

Chart 2: Variation in Average Temperature

Variation in Global Average Temperature (1900-2025)

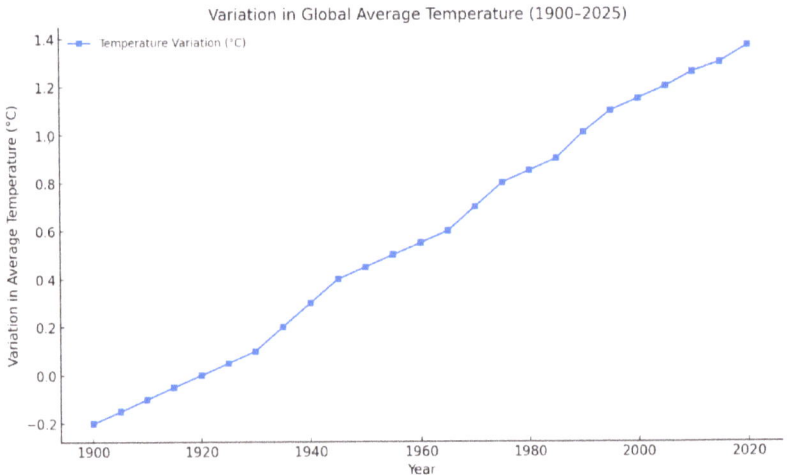

Source: Own elaboration based on the data presented in the Summary
Table at the end of this chapter

Moreover, the global dependence on oil has triggered geopolitical conflicts and energy crises that have impacted entire economies.

As the environmental impacts of fossil fuels became increasingly evident, the search for alternative energy sources gained momentum, paving the way for the development of renewable energies and nuclear power — topics we will explore in the chapters to come.

Chart 3: Correlation between CO_2 Emissions and Global Temperature

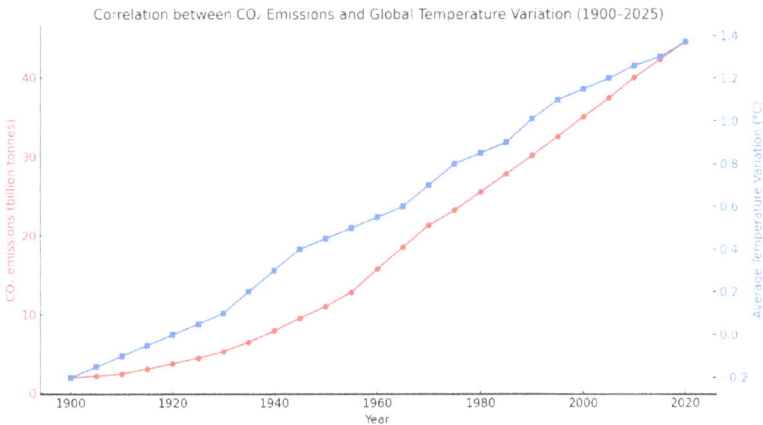

Correlation between CO₂ Emissions and Global Temperature Variation (1900-2025)

Source: Own elaboration based on the data presented in the Summary Table at the end of this chapter

The Search for More Efficient Energy Sources

Fossil fuels — such as coal, oil, and natural gas — have been fundamental to industrial and economic growth over the past centuries. However, their extraction, transportation, and consumption pose significant challenges. Beyond environmental impacts such as air pollutant emissions and potential oil spills, these resources play a crucial role in electricity costs and the economic stability of countries.

The volatility of oil and natural gas prices has direct effects on inflation and the competitiveness of industries. Dependence on fossil fuel imports creates geopolitical vulnerabilities, with exporting countries exerting economic influence over importers.

The True Impact of Renewable Energies

Although often promoted as clean energy sources, renewables also carry a considerable environmental footprint. The production of solar panels and wind turbines involves the extraction of rare minerals, such as neodymium and lithium, which have significant environmental impacts during their mining. Furthermore, the manufacturing, transportation, and installation of these systems generate carbon emissions.

Another critical factor is the need for supporting infrastructure. Since solar and wind energy are intermittent, they require the construction of backup systems such as large-scale batteries or reserve thermal power plants, often powered by natural gas. This increases the overall cost of electricity and reduces the actual environmental advantage of these technologies.

One of the significant limitations of renewable energy is its intermittency. Since the sun doesn't shine at night and the wind doesn't always blow with sufficient intensity, redundancy in electricity generation is necessary. This means that countries must maintain additional operational power plants to supply the grid during periods of low renewable output.

This redundancy comes with a significant cost to national power systems. The need to balance intermittent sources with stable ones — such as hydroelectric or thermal plants — increases investment in infrastructure and maintenance, which may be reflected in electricity prices for both consumers and industries.

While renewable energies play a fundamental role in the global energy mix, they are not, on their own, capable of providing a complete solution to humanity's energy needs. Their intermittency, environmental footprint, and need for support make it essential to combine them with other reliable energy sources. In this context, nuclear energy emerges as a crucial alternative to ensure clean, stable, and economically viable electricity — a topic we will explore in the chapters ahead.

Chart 4: Evolution of Renewable Energies

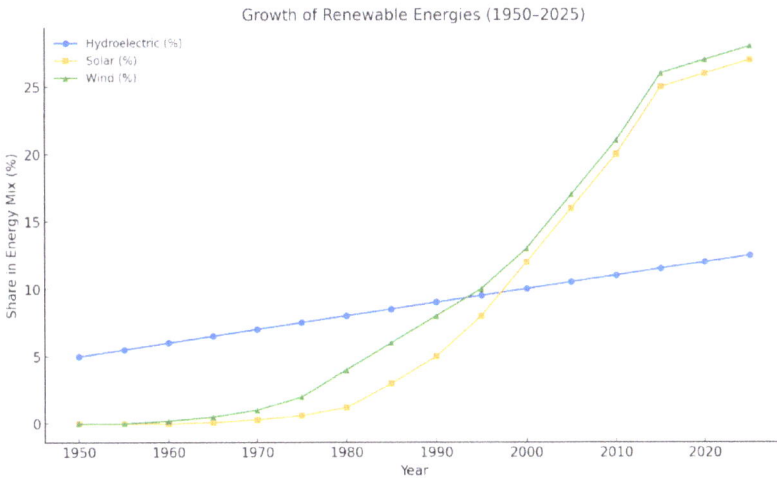

Growth of Renewable Energies (1950-2025)

Source: Own elaboration based on the data presented in the Summary Table at the end of this chapter

The Emergence of Nuclear Energy

The discovery of nuclear fission was one of the most critical milestones in modern science. In the early 20th century, physicists such as Henri Becquerel, Marie Curie, and Ernest Rutherford conducted pioneering experiments on radioactivity, revealing that certain elements emit energy spontaneously.

The real breakthrough, however, came in 1938, when German scientists Otto Hahn and Fritz Strassmann discovered that bombarding uranium with neutrons resulted in the splitting of the atomic nucleus, releasing a tremendous amount of energy. This process, later interpreted by Lise Meitner and Otto Frisch, was named "nuclear fission" and paved the way for a new energy era.

Nuclear fission proved to be an extremely efficient energy source. Unlike fossil fuels, which rely on chemical combustion and emit large amounts of carbon dioxide (CO_2), nuclear energy generates electricity without direct greenhouse gas emissions. Moreover, the energy density of uranium is incomparably higher than that of any known fuel: one kilogram of enriched uranium can generate millions of times more energy than one kilogram of coal or oil.

With this capacity, nuclear energy began to be seen as a viable solution to meet the growing global demand for electricity without exacerbating the environmental problems associated with traditional energy sources.

Chart 5: CO_2 Emissions by Different Energy Sources

Comparison of CO_2 Emissions from Different Energy Sources

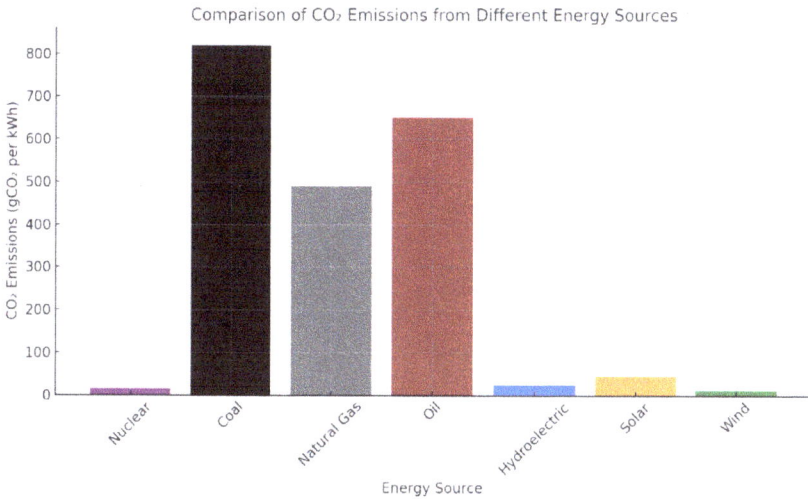

Source: Own elaboration based on the data presented in the Summary Table at the end of this chapter

Although the discovery of fission initially led to the development of nuclear weapons during World War II, the world soon began to explore the peaceful potential of this technology.

In 1951, the United States generated electricity from nuclear fission for the first time at the experimental EBR-I reactor in Idaho.

From that point on, several countries began investing in nuclear power plants to produce electricity. In 1954, the Soviet Union inaugurated the world's first commercial nuclear power plant in Obninsk. Since then, nuclear energy has become a fundamental part of the global energy matrix and today accounts for around 10% of the world's electricity.

Safety, efficiency, and a low carbon footprint have made nuclear energy one of the main alternatives for a sustainable energy future. In the next chapter, we will examine the central myths and realities surrounding this technology, exploring its peaceful applications and benefits across various sectors of the economy.

How Scientific Advances Led to the Discovery of Nuclear Fission

The theory behind nuclear fission is partly rooted in Albert Einstein's famous equation, $E=mc^2$, formulated in 1905. This equation establishes the relationship between energy (E) and mass (m), showing that a small amount of matter can be converted into a vast amount of energy when multiplied by the square of the speed of light (c). This principle was fundamental to understanding the immense energy potential contained within the atomic nucleus.

Although Einstein did not work directly on the discovery of nuclear fission, his theory of relativity — and his letters to U.S. President Franklin D. Roosevelt warning about the potential of uranium as an energy source — influenced the development of the first nuclear reactors. The peaceful application of this technology would come years later, as nuclear energy began to be explored as a viable alternative for electricity generation.

Chart 6: Energy Efficiency of Different Sources

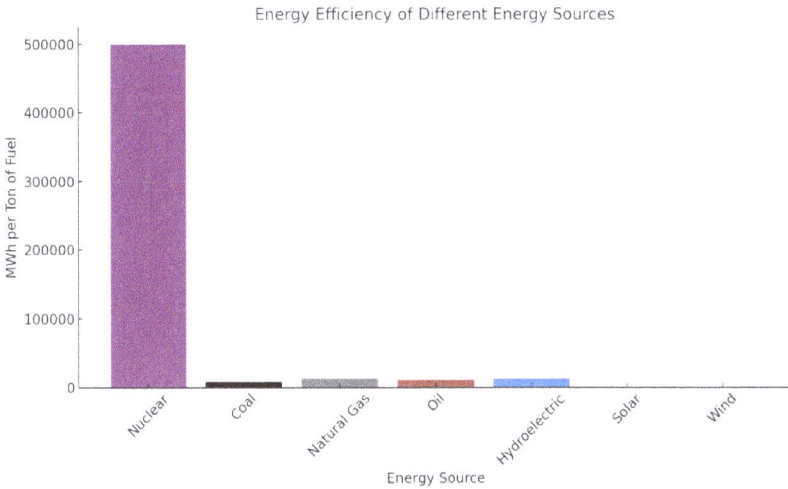

Energy Efficiency of Different Energy Sources

Source: Own elaboration based on the data presented in the Summary Table at the end of this chapter

Enriched uranium is a material — the "fuel" — essential to the operation of nuclear reactors.

It is uranium that has undergone an isotopic separation process to increase the concentration of the isotope U-235, which is more prone to fission compared to U-238, the most abundant isotope in nature. This enrichment process allows the chain reaction to occur in a controlled manner, releasing a significantly greater amount of energy than any conventional chemical reaction.

Comparison of Energy Intensity Among Different Energy Sources

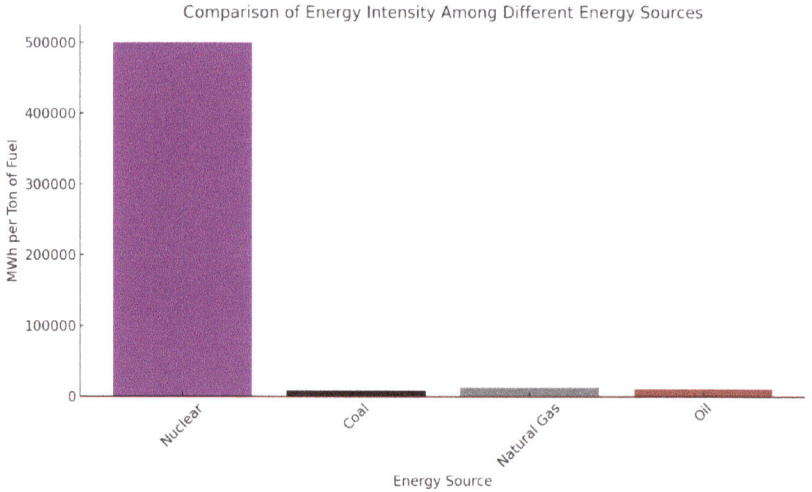

Source: Own elaboration based on the data presented in the Summary Table at the end of this chapter

The First Use of Nuclear Energy for Peaceful Purposes and Its Importance in the Global Energy Mix

Chart 8: Evolution of the Global Energy Mix

Evolution of Energy Sources in the Global Energy Mix (1950-2025)

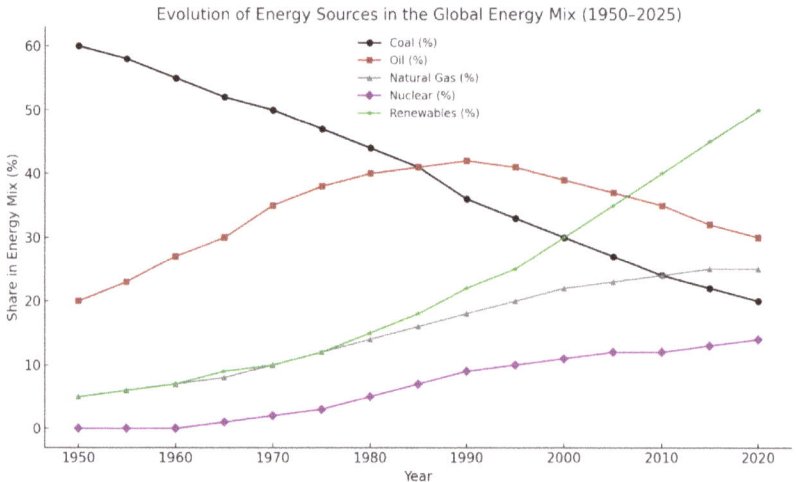

Source: Own elaboration based on the data presented in the Summary Table at the end of this chapter

Nuclear energy's safety, efficiency, and low carbon footprint have made it one of the leading alternatives for a sustainable energy future.

In the next chapter, we will explore the central myths and realities surrounding this technology, examining its peaceful applications and benefits across various sectors of the economy.

Nuclear energy stands out for its high energy efficiency and low carbon footprint when compared to other sources. One kilogram of enriched uranium can generate millions of times more energy than one kilogram of coal or oil. This energy density makes nuclear power an extremely competitive alternative in terms of cost-effectiveness and environmental impact.

When comparing the primary energy sources used globally, we observe striking differences:

- Coal: Although cheap and widely available, it is one of the most polluting sources, emitting considerable amounts of CO_2 and toxic particles.

- Oil and natural gas: These are versatile and used for both electricity and transportation, but they remain dependent on unstable markets and generate significant emissions.

- Renewable energies (solar and wind): They have a low carbon footprint, but face challenges related to intermittency, storage needs, and high infrastructure investment.

- Hydropower: A reliable renewable source, but its construction can cause significant environmental and social impacts due to the flooding of vast areas.

- Nuclear energy: Produces electricity continuously, with no direct carbon emissions and a low environmental impact during operation.

The Growth of Nuclear Energy and the Challenges Ahead

Currently, nuclear energy accounts for about 10% of global electricity production. Several countries continue to invest in this technology as an alternative to ensure energy security and reduce greenhouse gas emissions. China, Russia, France, and the United States are at the forefront of investments in new nuclear reactors, including initiatives focused on Small Modular Reactors (SMRs), which provide enhanced flexibility and reduced implementation costs.

Chart 9: Growth of Nuclear Energy by Continent

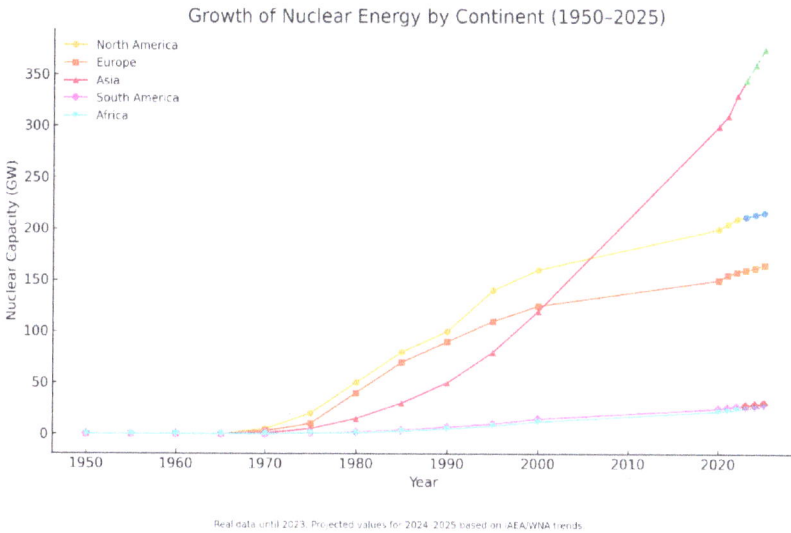

Growth of Nuclear Energy by Continent (1950-2025)

Real data until 2023. Projected values for 2024-2025 based on IAEA/WNA trends

Source: Own elaboration based on the data presented in the Summary Table at the end of this chapter

However, nuclear energy faces significant challenges:

– Public perception and fear of accidents: Incidents like Chernobyl and Fukushima have heightened concerns over nuclear safety despite significant technological advancements that have greatly improved the safety of modern reactors.

- Radioactive waste management: While the volume of nuclear waste is small compared to the total of toxic industrial waste, the storage and disposal of these materials require strict regulations.

- High initial costs: Building nuclear power plants requires significant investments, which are often only recovered in the long term.

- Political and regulatory challenges: Issues related to nuclear proliferation and differing national regulations affect the pace of nuclear energy expansion.

Chart 10: Nuclear Generation Capacity

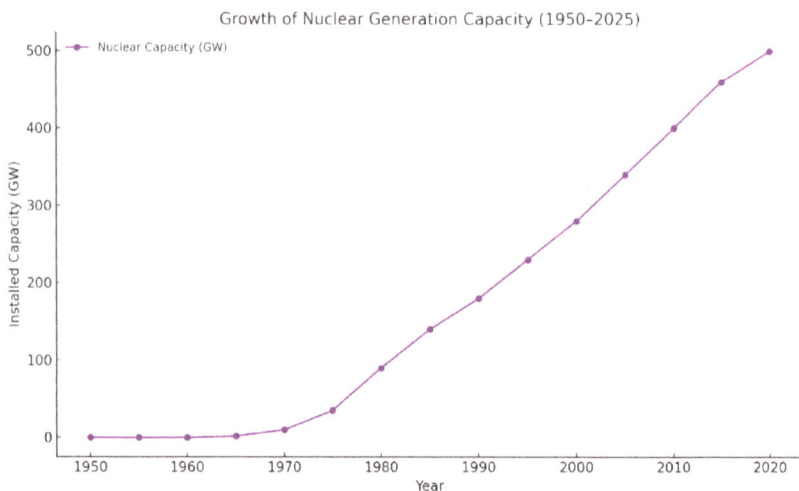

Growth of Nuclear Generation Capacity (1950-2025)

Source: Own elaboration based on the data presented in the Summary Table at the end of this chapter

The Need for a Balanced Approach to Energy Transition

In the face of 21st-century energy challenges, a balanced approach that combines different energy sources is essential to ensure a sustainable, secure, and affordable electricity supply.

Nuclear energy plays a crucial role in this balance, offering a low-carbon solution that can provide electricity continuously, regardless of weather conditions. Unlike intermittent sources such as solar and wind, nuclear power can operate continuously, ensuring stability for the electric grid.

Moreover, as new nuclear technologies — such as fourth-generation reactors and nuclear fusion — continue to develop, operational costs and safety concerns are expected to be further reduced.

The energy transition requires strategic planning that carefully weighs the strengths and limitations of each energy source. Overlooking nuclear power in the pursuit of a sustainable energy system could lead to higher costs for consumers and increased reliance on less stable and predictable sources.

Chart 11: Cost Comparison Among Various Energy Sources

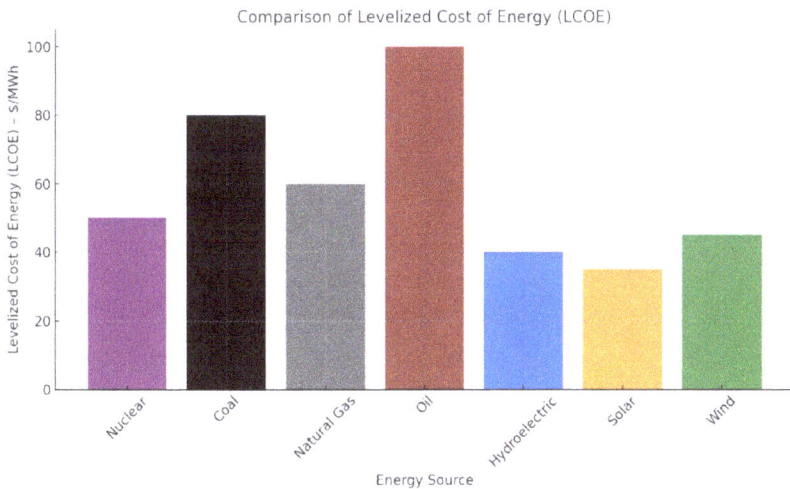

Source: Own elaboration based on the data presented in the Summary Table at the end of this chapter

Note: A detailed description of LCOE is provided in Chapter 6.

In the chapters ahead, we will delve deeper into the challenges and opportunities associated with nuclear energy, including issues of safety, waste management, and technological advancements.

Conclusion of the Present Chapter

Since the dawn of civilization, energy has been the driving force behind human development. The harnessing of fire, followed by the use of wood, coal, oil, and, more recently, electricity, has profoundly shaped the development of societies and driven remarkable technological advancement. Without a stable and accessible energy supply, industrialization would not have been possible, and the modern world as we know it would not exist.

Throughout history, energy sources have been progressively refined and adapted to meet the ever-growing needs of humanity. However, the choice of energy sources has a direct impact on the economy, national security, and the quality of life of populations. Any country needs to have a strategic and diversified energy plan to avoid excessive dependence and external vulnerabilities.

Nuclear energy emerges as a fundamental pillar of a balanced and sustainable energy system. Despite the controversies and challenges, it remains one of the few sources capable of supplying electricity continuously, regardless of weather conditions, and with an extremely low environmental impact in terms of carbon emissions.

Excluding nuclear energy from a country's energy matrix is a grave strategic mistake that directly affects electricity costs for both consumers and industries. Countries that have phased out nuclear power, such as Germany, have experienced substantial increases in electricity prices and a greater reliance on natural gas imports, leaving them more vulnerable to geopolitical crises.

In addition, an overreliance on intermittent sources like solar and wind—without a stable foundation of constant generation—compels governments to maintain backup thermal power plants or rely on electricity imports from neighboring countries, potentially resulting in energy instability and increased costs. Nuclear energy avoids this scenario by ensuring a predictable and stable supply, reducing both economic and political risks.

Chart 12: Ideal Energy Mix

Ideal Energy Mix for a Sustainable and Secure Country

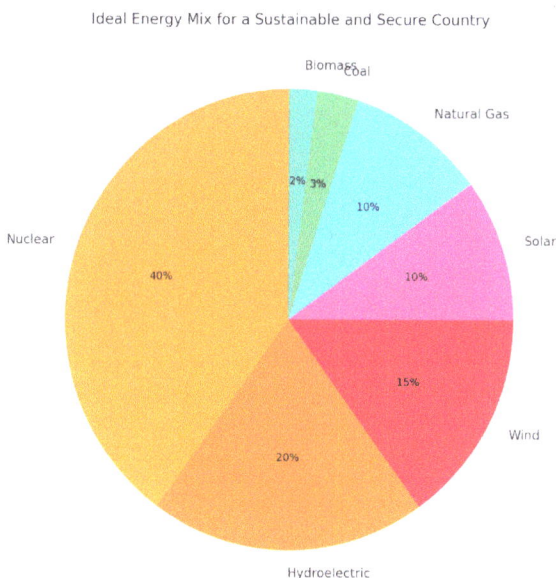

Source: Own elaboration based on the data presented in the Summary Table at the end of this chapter

Here is a chart representing an ideal energy mix for a country aiming to ensure a stable supply, minimize environmental impact, and maintain competitive prices for consumers and industries.

Structure of the Ideal Energy Mix:

- Nuclear (40%): Solid base for continuous generation, low carbon emissions, and grid stability.

- Hydropower (20%): A reliable renewable source with the ability to quickly adjust electricity supply.

- Wind (15%) and Solar (10%): Complement the mix with clean energy sources, although they require backup due to their intermittent nature.

- Natural Gas (10%): A supporting source for peak demand, less polluting than coal.

- Coal (3%) and Biomass (2%): Reserved for strategic purposes and specific needs.

Table 1: Sources Consulted in Chapter 1

Description
Our World in Data – Energy Production and Consumption Global historical energy production data.
International Energy Agency (IEA) – World Energy Outlook 2023 Analysis and projections on global energy markets.
U.S. Energy Information Administration (EIA) – Energy Explained Technical explanation of energy sources and their use.
World Nuclear Association – Nuclear Power Facts Statistics and insights on nuclear energy.
IPCC – Sixth Assessment Report Impacts of Fossil Fuels and Energy on Climate Change.
MIT Energy Initiative – The Future of Nuclear Energy Nuclear technology analysis and future scenarios.
British Petroleum – Statistical Review of World Energy 2022 Global energy data by source and region.
World Bank – Energy Use Indicators Economic and energy development indicators.
International Renewable Energy Agency (IRENA) – Renewable Capacity Statistics Growth of renewables and investment trends.
European Commission – Energy and Climate Policy European energy transition strategy.
National Renewable Energy Laboratory (NREL) – Cost and Efficiency Reports Reports on efficiency and cost of renewable energies.
United Nations – Sustainable Energy for All UN program to promote access to clean energy.

Preparation for the Next Chapter: The History and Evolution of Nuclear Energy

The next chapter will explore the evolution of nuclear energy from its earliest discoveries to its modern development. We will discuss how different countries adopted this technology, the main advancements achieved over the years, and how innovation is shaping the future of nuclear power. Additionally, we will examine the challenges and solutions aimed at making nuclear energy safer and more efficient.

Humanity's journey in search of reliable and sustainable energy sources continues. When properly utilized and regulated, nuclear energy can be one of the most effective solutions to the energy challenges of the 21st century.

Chapter 2 –
The History and Evolution of Nuclear Energy:
From Discovery to the Modern Era

Nuclear energy is undoubtedly one of the most impactful discoveries of modern history. Its development has shaped geopolitics, driven scientific advances, and redefined the concept of energy security. However, to understand its current significance and future prospects, it is essential to return to its origins and explore how this form of energy was discovered, developed, and applied over time.

From the first studies on radioactivity in the late 19th century to today's cutting-edge nuclear reactors, the journey of nuclear energy has been marked by decisive moments. Revolutionary scientific breakthroughs, such as nuclear fission, gave rise to technologies of enormous potential—but also to ethical and geopolitical dilemmas. World War II marked a turning point, transforming nuclear knowledge into an unprecedented instrument of destruction, leading to the development of the atomic bombs dropped on Hiroshima and Nagasaki.

However, right after the horrors of war, a new chapter began: the use of nuclear energy for peaceful purposes. The first nuclear power plants emerged in the 1950s, promising an energy revolution based on a highly efficient and virtually carbon-free fuel. In the following years, countries around the world began investing in nuclear energy to meet their growing electricity demands.

Yet, the trajectory of this technology has not been linear. Periods of great optimism were followed by moments of skepticism and fear. Accidents such as those at Three Mile Island (1979), Chernobyl (1986), and Fukushima (2011) deeply impacted public opinion and led several nations to reconsider their nuclear programs. While some countries chose to phase out nuclear power, others continued to invest in reactor modernization and technological safety.

Today, nuclear energy is once again at the center of global debates on energy transition and the decarbonization of the global economy. With the growing need for reliable and sustainable energy sources, new nuclear technologies are being developed, such as small modular reactors (SMRs) and advances in nuclear fusion, which could completely reshape the energy sector in the coming decades.

In this chapter, we will explore the history and evolution of nuclear energy—from its early scientific discoveries to its impact on geopolitics and the global energy landscape. Understanding this journey is crucial for assessing the challenges and opportunities of the present and for anticipating the future trajectory of this technology.

The Discovery of Nuclear Fission (1938)

The discovery of nuclear fission was one of the most critical moments in the history of science, as it revealed the possibility of releasing immense amounts of energy from the nucleus of the atom. This breakthrough was the result of experiments conducted by Otto Hahn and Fritz Strassmann, who demonstrated that the nucleus of uranium could be split into smaller fragments.

However, the interpretation of this phenomenon was not immediate. Lise Meitner and her nephew Otto Frisch were responsible for providing the theoretical explanation for nuclear fission, demonstrating that the process released an extraordinary amount of energy, as predicted by Einstein's famous equation $E=mc^2$.

The impact of this discovery was immense. Scientists quickly realized that nuclear fission could be used to generate energy on an industrial scale but also to develop weapons of mass destruction. The race to harness the power of the atom had only just begun.

By the late 1930s, nuclear physics was rapidly expanding. Scientists had already discovered that bombarding certain elements with neutrons was possible to induce nuclear reactions. Uranium, one of the heaviest naturally occurring elements, was a primary target of these experiments.

In December 1938, German chemists Otto Hahn and Fritz Strassmann, working at the Kaiser Wilhelm Institute, conducted an experiment that would forever change the understanding of atomic structure. They bombarded **uranium-235** atoms with neutrons and expected to obtain slightly heavier elements.

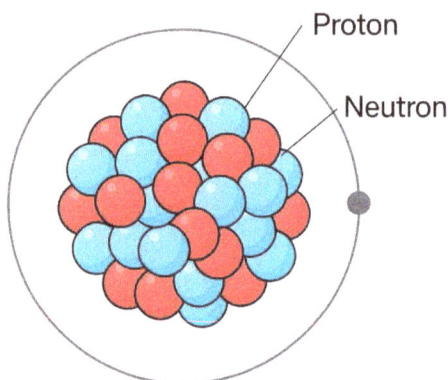

Diagram of the Uranium-235 Atom –
showing the nucleus structure with
protons and neutrons

However, upon analyzing the products of the reaction, they found barium—a much lighter element than uranium. This result was unexpected and contradicted everything known at the time about the structure of atomic nuclei. How could such a large and heavy atom like uranium split into two smaller fragments?

Otto Hahn and Fritz Strassmann initially did not understand what had happened. They reported their findings, but it was Lise Meitner who realized the true nature of the phenomenon.

Lise Meitner was a brilliant Austrian physicist who had worked alongside Otto Hahn for many years. However, due to the Nazi persecution of Jews, she was forced to flee Germany in 1938, eventually settling in Sweden. Even far from the laboratory,

Meitner maintained communication with Hahn and received experimental data on the uranium reaction.

Together with her nephew, physicist Otto Frisch, Meitner analyzed the results and concluded that the only possible explanation was that the uranium nucleus had split into two, releasing a tremendous amount of energy in the process.

They realized that this phenomenon aligned with Einstein's **equation (E=mc^2)**: a small portion of the nucleus's mass was being converted into an enormous amount of energy when subjected to an extremely high velocity.

Otto Frisch coined the term "nuclear fission" for the process, drawing an analogy to biological cell division. This explanation was published in early 1939 and soon attracted the attention of the international scientific community.

Diagram of Nuclear Fission

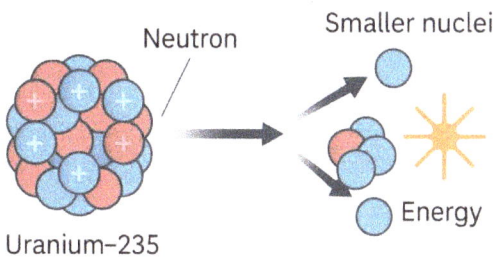

Simplified example of the fission reaction of Uranium-235 after neutron absorption.

Table 2: Nuclear Fission Reaction (U-235)

Before Fission	After Fission
U-235 + neutron	Ba-141 + Kr-92 + 3 neutrons + energy (~200 MeV)

Source: Own elaboration based on the data presented in the Summary Table at the end of this chapter

Early Discoveries (Late 19th – Early 20th Century)

The discovery of radioactivity at the end of the 19th century was a fundamental milestone in modern science, paving the way for the development of nuclear energy. Until then, the structure of the atom remained a mystery, and scientists believed it to be the smallest indivisible unit of matter. However, pioneering experiments showed that atoms were far more complex than previously imagined.

Advancements in this field occurred in three crucial stages: the discovery of radioactivity by Henri Becquerel, the research conducted by Marie and Pierre Curie, and the development of the first atomic models by Ernest Rutherford and Niels Bohr. These scientists were responsible for uncovering the mysteries of the atom and laying the foundation for the use of nuclear energy in the 20th century.

Henri Becquerel, a French physicist and specialist in fluorescence, conducted an experiment in 1896 that would change the course of science. He was studying the effects of sunlight on fluorescent materials and decided to test whether uranium salts could spontaneously emit X-rays.

Becquerel placed a uranium crystal on a photographic plate wrapped in black paper and left the sample exposed to sunlight, expecting the material to absorb the light's energy and reemit it as radiation. However, due to several cloudy days, the experiment was interrupted, and the plates remained in the dark. To his surprise, when he developed the plates days later, he found that the image of the uranium crystal was clearly imprinted on them—even without exposure to the sun.

This meant that uranium was emitting an unknown form of radiation spontaneously and continuously without any external energy source. Becquerel had just discovered natural radioactivity[1]—a phenomenon that contradicted existing

---

[1] This phenomenon is now known as NORMs. NORM stands for Naturally Occurring Radioactive Material. It refers to materials that contain naturally occurring radionuclides such as uranium (U), thorium (Th), radium (Ra), and potassium-40 (K-40), which are found in the Earth's crust. These materials are encountered in various industries, particularly those involving the extraction of natural resources, including:

• Oil and gas: underground deposits may contain radionuclides that accumulate in equipment and pipelines.

• Mining: ores such as phosphate, coal, and uranium contain varying amounts of radioactive material.

• Fertilizer industry: phosphates often contain natural radionuclides.

• Groundwater: may contain dissolved radioactive isotopes such as radium-226 and radium-228.

scientific understanding and suggested that atoms were not as stable as previously believed.

Inspired by Becquerel's experiments, the scientist couple Marie and Pierre Curie decided to further investigate radioactivity. In 1898, while working with tons of uranium ore (pitchblende), the Curies identified two new highly radioactive elements: **polonium** (named in honor of Marie Curie's native Poland) and **radium**.

The discovery of radium was revolutionary. The element emitted intense, continuous radiation without needing to be activated by heat or light. This phenomenon challenged classical theories of matter and revealed a new source of energy stored within the atomic nucleus.

Marie Curie, in addition to being the first woman to win a Nobel Prize, was a pioneer in the study of radioactivity—a field she helped coin. Her work was crucial for the development of medical applications of radiation, including radiotherapy for cancer treatment. However, constant handling of radioactive materials without protection severely affected her health, leading to her death in 1934 from a radiation-induced illness.

The Curies' discoveries showed that certain elements had the ability to spontaneously emit particles and energy. This realization raised fundamental questions about atomic

––––––––––––––––––––––––

There is also the acronym TENORM (Technologically Enhanced Naturally Occurring Radioactive Material), which refers to NORM materials whose levels of radioactivity have been increased by industrial processes.

structure and led to the development of new scientific models to explain these phenomena.

As studies on radioactivity progressed, it became clear that the traditional atomic model was insufficient to explain the newly observed behaviors. This led to the creation of more sophisticated models, the most influential being those proposed by Ernest Rutherford and Niels Bohr.

In 1911, the New Zealand physicist Ernest Rutherford conducted a famous experiment known as the Gold Foil Experiment. He bombarded a thin gold foil with alpha particles (helium nuclei) and observed that while most particles passed through the foil undisturbed, some were deflected at unexpected angles. This led him to conclude that atoms were not solid, homogeneous structures as previously thought but rather composed of a small, dense, positively charged nucleus containing almost all of the atom's mass, surrounded by a vast region of empty space in which electrons orbited.

This model revolutionized physics by demonstrating that matter is essentially empty space and that the energy released in radioactivity originates from the atomic nucleus.

Ernest Ru[...]
a Gold Foil

Atomic nucleus
alpha prarticles

Electons ion radioaive angles.

Experiment

Two years later, in 1913, Danish physicist Niels Bohr refined Rutherford's model by proposing that electrons orbit the nucleus in discrete energy levels, like concentric shells around the nucleus. This model explained why atoms emit and absorb light only at specific wavelengths—a phenomenon fundamental to understanding spectroscopy and, later, quantum mechanics.

Bohr's model was a crucial advancement for nuclear science, as it helped explain how atoms interact with energy and how processes such as nuclear fission occur, in which atomic nuclei split, releasing a tremendous amount of energy.

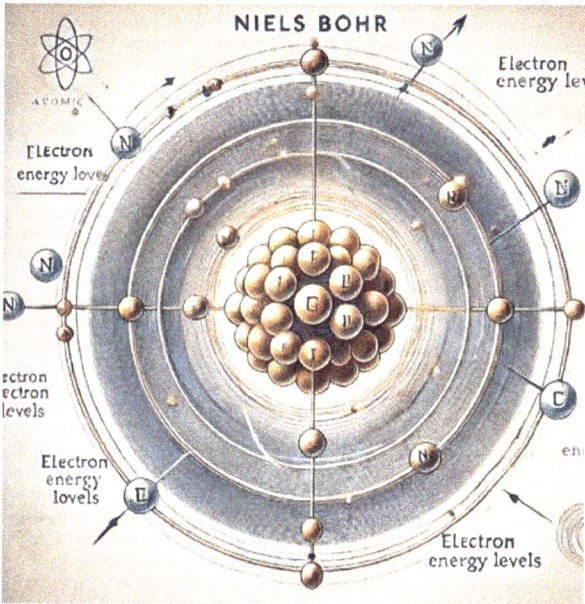

The turn of the 19th to the 20th century marked the beginning of a new scientific era. The discovery of radioactivity by Henri Becquerel, the groundbreaking work of the Curies, and the atomic models proposed by Rutherford and Bohr transformed our understanding of matter and energy.

These advances were fundamental to the development of nuclear energy, as they demonstrated that the atomic nucleus stored immense amounts of energy. This knowledge would pave the way for the discovery of nuclear fission in the 1930s—an event that would change the world forever.

Early Speculations and the Military Potential of Nuclear Fission

The discovery of nuclear fission in late 1938 opened a vast field of scientific, industrial, and military possibilities. The fact that a single uranium atom could release an immense amount of energy generated great excitement but also raised ethical, political, and economic questions about the use of this new technology.

The scientific community quickly realized that fission could be explored in two main ways:

- For the generation of electricity through the construction of nuclear reactors capable of producing energy efficiently and continuously.

- For the creation of nuclear weapons, using the energy of fission to unleash explosions of unprecedented destructive power.

Discussions about these uses began almost immediately after the publication of Lise Meitner and Otto Frisch's findings in 1939. Different groups of scientists and governments reacted differently ranging from optimism about a promising energy future to fears that humanity was opening a "Pandora's box."

The idea of using nuclear fission to generate electricity was seen as revolutionary. Scientists understood that if nuclear reactions could be controlled safely and continuously, this technology could provide a clean, powerful, and virtually inexhaustible energy source.

The theoretical advantages of nuclear energy were clear:

- High energy density – A small volume of uranium could yield far more energy than any fossil fuel.

- Low carbon emissions – Unlike coal and oil, nuclear fission emitted no CO_2, making the technology promising for a growing industrialized world.

- Energy independence – Countries lacking abundant coal or oil reserves could use nuclear energy to meet their energy needs.

Despite these promises, building a functional nuclear reactor posed significant challenges. To ensure that fission occurred in a controlled (non-explosive) way—as opposed to the uncontrolled release in a nuclear bomb—a system had to be designed to regulate the chain reaction, ensuring stability and safety.

In 1942, physicist Enrico Fermi, working in the United States, led the construction of the first experimental nuclear reactor, known as Chicago Pile-1. This was the first concrete step in demonstrating that nuclear energy could be used for peaceful purposes. However, while some scientists explored the peaceful applications of fission, others focused on a far more destructive use: the development of nuclear weapons.

The Military Potential: The Road to the Atomic Bomb

If nuclear fission could be used to safely and continuously generate electricity, it could also be harnessed to create massive explosions. By mid-1939, the possibility of an atomic bomb was already a serious concern within the scientific community.

Physicists such as Albert Einstein and Leó Szilárd were alarmed at the idea that Nazi Germany might develop a nuclear weapon before the Allies. It was in this context that Einstein signed the famous letter to President Franklin D. Roosevelt, warning about the military potential of fission and encouraging the United States to begin research into nuclear weapons.

This led to the creation of the Manhattan Project—a top-secret scientific and military effort that would culminate just a few years later in the bombings of Hiroshima and Nagasaki in 1945.

This duality of nuclear energy—as a source of hope and progress but also as a tool of mass destruction—would profoundly shape the debates surrounding nuclear technology in the decades to follow.

The Initial Controversies: Scientific Debate

From the earliest months after the discovery of fission, there was intense debate among scientists over which applications should be prioritized. Some of the main controversies included:

- **The ethics of developing nuclear weapons** – Many physicists were reluctant to support the military use of nuclear fission, fearing long-term consequences. Some, like Niels Bohr, argued that nuclear knowledge should be shared globally to prevent an arms race.

- **The risk of nuclear accidents** – Early on, scientists such as Enrico Fermi warned that if the chain reaction was not adequately controlled, it could lead to uncontrolled explosions or radioactive contamination, creating technical challenges for building safe reactors.

- **The geopolitical impact of nuclear energy** – There were concerns about how nuclear technology might shift the global balance of power. Over time, this led to international tensions and the need for nuclear control treaties.

The impact of these discussions was enormous. While some scientists remained committed to developing nuclear energy for peaceful use, others withdrew entirely from the field, troubled by the political and military implications of the technology.

In conclusion, the discovery of nuclear fission quickly split the scientific community and governments between two opposing paths: the use of fission for energy generation and the development of nuclear weapons.

On the one hand, fission promised to meet global energy demand with a powerful and clean energy source; on the other, the potential to create weapons with unprecedented destructive power radically changed the political and military landscape of the time.

The quest to balance peaceful and military uses of nuclear fission would shape the course of history in the 20th century— and continue to spark debate to this day.

In the following chapters, we will explore how these choices led to the creation of the first nuclear reactors and the development of atomic bombs, marking the beginning of the Nuclear Age.

World War II and the Manhattan Project

World War II (1939–1945) was one of the most turbulent periods in human history, and nuclear energy played a crucial role in that conflict. The discovery of nuclear fission in 1938 raised questions about its energy potential; however, it was the military use of this technology that altered the course of the war and global geopolitics.

The possibility of building a nuclear weapon based on atomic fission drew great interest among scientists and military leaders. The fear that Nazi Germany was developing an atomic bomb led the United States to mobilize its leading scientists for the Manhattan Project—an unprecedented scientific and industrial effort to create the first nuclear bombs.

The results of this project became tragically evident on August 6 and 9, 1945, when the Japanese cities of Hiroshima and Nagasaki were destroyed by atomic bombs, demonstrating to the world the devastating power of nuclear energy.

In this section, we will analyze how advances in nuclear physics were used for military purposes, the fundamental role of Albert Einstein and his letter to Roosevelt, the development of the atomic bomb in the U.S., and the impact of the Hiroshima and Nagasaki bombings.

How Advances in Nuclear Physics Were Used for Military Purposes

The discovery of nuclear fission in 1938 was a scientific milestone, but it also raised concerns about its military applications. Scientists realized that by inducing a chain fission reaction in a sufficient quantity of uranium-235 or plutonium-239, it was possible to release colossal energy in the form of an explosion.

The theory was simple, but the engineering required to build a nuclear weapon posed complex challenges. The main issue was obtaining enough fissile material to sustain an explosive chain reaction. Only two known substances could be used:

- **Uranium-235** (U-235) – a rare isotope of natural uranium that needs to be enriched.

- **Plutonium-239** (Pu-239) – artificially produced from uranium-238 in nuclear reactors.

Governments quickly understood that whoever mastered this technology would hold the strategic advantage on the battlefield. The race to develop the atomic bomb had begun, and Nazi Germany was seen as the primary threat.

Albert Einstein and the Letter to Roosevelt

In 1939, physicists exiled from Germany, such as Leó Szilárd and Edward Teller, were deeply concerned about the possibility that Adolf Hitler might obtain a nuclear weapon before the Allies. They knew that German scientists were studying nuclear fission and feared that the Nazi regime might attempt to use it militarily.

To alert the U.S. government, Szilárd persuaded Albert Einstein to sign a letter addressed to President Franklin D. Roosevelt. This letter, sent on **August 2, 1939**, explained the discovery of nuclear fission, and warned that Germany might be working to develop an atomic bomb.

Albert Einstein
Old Grove Rd.
Nassau Point
Peconic, Long Island
August 2nd, 1939

F.D. Roosevelt,
President of the United States,
White House
Washington, D.C.

Sir:

Some recent work by E. Fermi and L. Szilard, which has been communicated to me in manuscript, leads me to expect that the element uranium may be turned into a new and important source of energy in the immediate future. Certain aspects of the situation which has arisen seem to call for watchfulness and, if necessary, quick action on the part of the Administration. I believe therefore that it is my duty to bring to your attention the following facts and recommendations:

In the course of the last four months it has been made probable - through the work of Joliot in France as well as Fermi and Szilard in America - that it may become possible to set up a nuclear chain reaction in a large mass of uranium, by which vast amounts of power and large quantities of new radium-like elements would be generated. Now it appears almost certain that this could be achieved in the immediate future.

This new phenomenon would also lead to the construction of bombs, and it is conceivable - though much less certain - that extremely powerful bombs of a new type may thus be constructed. A single bomb of this type, carried by boat and exploded in a port, might very well destroy the whole port together with some of the surrounding territory. However, such bombs might very well prove to be too heavy for transportation by air.

-2-

The United States has only very poor ores of uranium in moderate quantities. There is some good ore in Canada and the former Czechoslovakia, while the most important source of uranium is Belgian Congo.

In view of this situation you may think it desirable to have some permanent contact maintained between the Administration and the group of physicists working on chain reactions in America. One possible way of achieving this might be for you to entrust with this task a person who has your confidence and who could perhaps serve in an inofficial capacity. His task might comprise the following:

a) to approach Government Departments, keep them informed of the further development, and put forward recommendations for Government action, giving particular attention to the problem of securing a supply of uranium ore for the United States;

b) to speed up the experimental work, which is at present being carried on within the limits of the budgets of University laboratories, by providing funds, if such funds be required, through his contacts with private persons who are willing to make contributions for this cause, and perhaps also by obtaining the co-operation of industrial laboratories which have the necessary equipment.

I understand that Germany has actually stopped the sale of uranium from the Czechoslovakian mines which she has taken over. That she should have taken such early action might perhaps be understood on the ground that the son of the German Under-Secretary of State, von Weizsäcker, is attached to the Kaiser-Wilhelm-Institut in Berlin where some of the American work on uranium is now being repeated.

Yours very truly,
(Albert Einstein)

The letter had a tremendous impact. Roosevelt created a committee to investigate the feasibility of a nuclear weapon, which became the embryo of what would later become the Manhattan Project.

It is important to note that Einstein did not participate directly in the development of the bomb. After the war, he regretted his indirect involvement and became one of the leading advocates for nuclear disarmament.

> **Contextual Note:**
> This letter was written on August 2, 1939. World War II would only begin a month later, on September 1, with the invasion of Poland by Nazi Germany. The United States would formally enter the conflict in December 1941, following the attack on Pearl Harbor.
>
> Even so, Einstein's letter, drafted with the help of Leó Szilárd, demonstrates a remarkable ability to foresee the scientific and geopolitical dangers ahead. More than a warning, it was a call to responsibility.
>
> The real merit, however, was also in Roosevelt, who chose to listen to science rather than reject it, launching one of the most controversial and transformative projects of the 20th century — the Manhattan Project.

The Development of the Atomic Bomb in the U.S. – The Manhattan Project

Faced with fears that the Nazis might succeed in building a bomb before the Allies, the United States decided to invest heavily in the development of this technology.

In 1942, the U.S. government launched the Manhattan Project, an ultra-secret operation to design and build the atomic bomb. The project involved:

- Over 130,000 people, including scientists, engineers, and military personnel.

- An estimated cost of 2 billion dollars at the time (equivalent to tens of billions today).

-Three major research centers:

- **Los Alamos (New Mexico)** – the central laboratory for bomb design, under the leadership of J. Robert Oppenheimer.

- **Oak Ridge (Tennessee)** – responsible for uranium enrichment.

- **Hanford (Washington)** – production of plutonium-239 for the bomb.

Two types of bombs were developed:

- *Little Boy* (dropped on Hiroshima) – used uranium-235 and worked through a "gun-type" firing mechanism.

- *Fat Man* (dropped on Nagasaki) – used plutonium-239 and relied on an implosion method to reach critical mass.

Before being used in war, the bombs were tested in the New Mexico desert on July 16, 1945, during the Trinity test—the first nuclear explosion in history.

Table 3: Comparison Between U-235 and Pu-239

Parameter	U-235	Pu-239
Origin	Natural (0.7% of uranium)	Produced in reactors from U-238
Main Use	Civil reactors and weapons (Little Boy)	Nuclear weapons (Fat Man)
Acquisition Difficulty	Requires enrichment	Requires reactor and chemical separation

Source: Own elaboration based on the data presented in the Summary Table at the end of this chapter

Hiroshima and Nagasaki: The Impact of the First Nuclear War

Even after Germany's surrender in May 1945, Japan refused to surrender unconditionally. The United States, fearing that a conventional invasion of Japan would result in millions of casualties, decided to use the atomic bomb to force Japan's capitulation.

Hiroshima – August 6, 1945

The Little Boy bomb was dropped on the city of Hiroshima at 8:15 a.m.

The explosion generated heat exceeding one million degrees Celsius at the epicenter.

It is estimated that 70,000 to 80,000 people died instantly, and tens of thousands more succumbed in the following months due to radiation exposure.

"Little Boy' Bomb
The bomb dropped on
iroshima on August 6, 1945

Enola Gay Aircraft
The B-29 bomber that tro-
pped the bomb on Hiroshima

Nagasaki – August 9, 1945

The Fat Man bomb was dropped on Nagasaki at 11:02 a.m.

The city's mountainous topography reduced the damage, but still, around 40,000 people died instantly, and more than 70,000 perished in the months that followed.

"Fat Man' Bomb: The bomb dropped on Nagasaki on August 9, 1945. Bockscar Aircraft: The B-29 bomber that dropped the bomb on Nagasaki.

Table 4: Comparison Between Nuclear Bombs: Little Boy vs Fat Man

Characteristic	Little Boy	Fat Man
Fissile Material	U-235	Pu-239
Mechanism	Gun-type	Implosion
Target City	Hiroshima	Nagasaki

Source: Own elaboration based on the data presented in the Summary
Table at the end of this chapter

Three days after the bombing of Nagasaki, Japan officially surrendered on **August 15, 1945**, bringing World War II to an end.

In conclusion, the use of nuclear energy during World War II changed the world forever. The destructive power demonstrated in Hiroshima and Nagasaki created a **nuclear paradox**: on the one hand, nuclear fission could be used for peaceful purposes, but on the other, its destructive potential introduced a new kind of global threat.

From that moment on, nuclear energy would become a geopolitical weapon, marking the beginning of the Cold War and the arms race. At the same time, scientists began exploring ways to harness this technology for energy production, marking the beginning of the nuclear power plant era.

In the following chapters, we will examine how the world responded to this new reality, seeking a balance between the benefits of nuclear energy and the risks of atomic weapons.

The Era of Atoms for Peace (1945–1970)

After the bombings of Hiroshima and Nagasaki in 1945, nuclear energy came to be viewed with great fear. The destruction caused by the atomic bombs demonstrated the devastating power of nuclear fission and led to intense international pressure to regulate the use of this technology.

However, soon after the war, scientists and governments began exploring the peaceful use of nuclear energy, aiming to harness its enormous potential for electricity generation and other industrial and medical applications.

This period was marked by an effort to reshape the image of nuclear energy, highlighting its role as a solution to the growing global energy demand. This new phase was symbolized by the "Atoms for Peace" program, announced by U.S. President Dwight D. Eisenhower in 1953. At the same time, the Soviet Union and other countries also launched programs to develop nuclear energy for civilian purposes.

In this chapter, we will examine how nuclear energy was promoted for peaceful purposes, the development of the first nuclear power plants, and the establishment of the International Atomic Energy Agency (IAEA).

"Atoms for Peace" and Eisenhower's 1953 Speech

After World War II, the United States and the Soviet Union entered the Cold War—an ideological and technological rivalry that included the nuclear arms race. However, even as both countries expanded their arsenals, there was growing concern over the risks of nuclear proliferation.

On December 8, 1953, President Dwight D. Eisenhower addressed the United Nations General Assembly, proposing the creation of the "Atoms for Peace" program. This initiative had the following objectives:

- Promote the peaceful use of nuclear energy, especially in electricity generation.

- Establish international agreements to prevent nuclear technology from being used for military purposes.

- Provide assistance to other countries to develop their own peaceful nuclear programs under international supervision.

Eisenhower's speech was a milestone in nuclear diplomacy and helped consolidate the idea that atomic energy could be a tool for progress—not just a threat.

The First Nuclear Power Plants for Electricity Generation

A few years after World War II, advancements in nuclear technology made it possible to build the first nuclear power plants for electricity generation. Two milestones stand out during this period:

Obninsk (USSR, 1954) – The World's First Nuclear Power Plant

The Soviet Union was the first country to inaugurate a nuclear power plant connected to the national electrical grid. The Obninsk plant, located about 110 km from Moscow, began operation on June 27, 1954.

Power output: 5 megawatts of electricity (MW).

Objective: Demonstrate the viability of nuclear energy as a source of electricity.

Technology: Graphite-moderated, light-water-cooled reactor— similar to the future Soviet RBMK reactors.

Although its capacity was small, Obninsk proved that nuclear energy was feasible for civilian use, marking the beginning of a new era in electricity production.

Obninsk Nuclear Power Plant

Shippingport (USA, 1957) – The First Large-Scale Commercial Nuclear Power Plant.

The United States inaugurated its first commercial nuclear power plant, Shippingport, on December 23, 1957. Unlike Obninsk, which was still experimental in nature, Shippingport was built to provide electricity continuously and reliably.

Power output: 60 megawatts of electricity (MW), significantly higher than Obninsk's capacity.

Technology: Pressurized Water Reactor (PWR), a model that would become the most widely used in the world.

Significance: It proved that nuclear energy was commercially viable and competitive with other energy sources.

With the success of these early plants, several countries began investing in nuclear energy, marking the start of a global expansion of the technology.

Shippingport Nuclear Power Plant

Global Expansion of Nuclear Energy

Following the first successful initiatives in the peaceful use of nuclear energy, such as the Obninsk and Shippingport power plants, several countries began investing in the technology. Several factors drove the global expansion of nuclear energy between the 1950s and 1970s:

- Need for Energy Diversification: Countries sought to reduce dependence on fossil fuels and ensure energy security.

- Technological Advancements: Development of more efficient and safer reactors.

- International Prestige: Possessing nuclear technology was seen as a symbol of development and technological power.

Main Reactor Types Developed

During this period, two main types of nuclear reactors stood out:

Graphite-Moderated, Light-Water-Cooled Reactor

Function: Uses graphite as a moderator to slow down neutrons and light water (H_2O) as a coolant to remove the heat generated in nuclear fission.

Advantages: Allows the use of natural uranium as fuel, reducing the need for enrichment.

Disadvantages: Larger volume and greater structural complexity.

Diagram of a Graphite-Moderated, Light-Water-Cooled Reactor

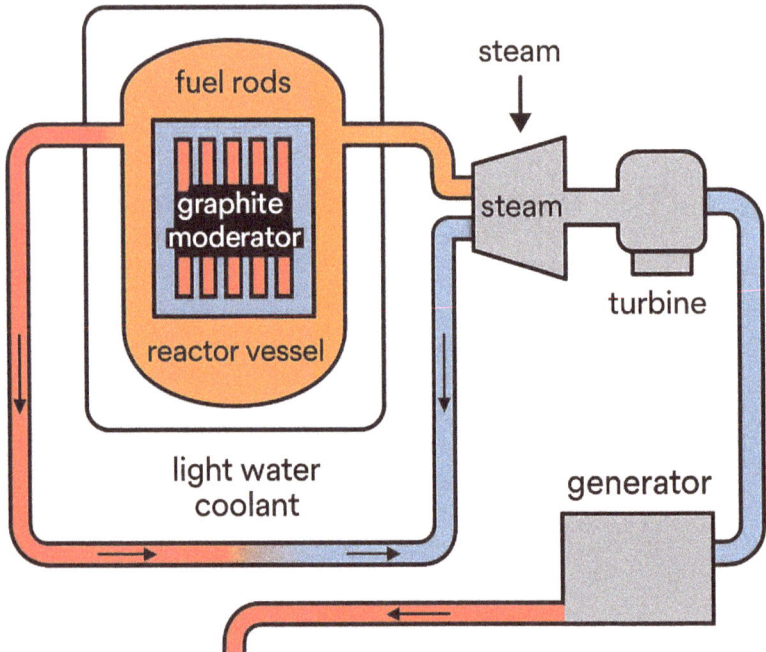

Pressurized Water Reactor (PWR):

Function: Uses light water both as a moderator and as a coolant. The water is kept under high pressure to prevent it from boiling inside the reactor, where it transfers the heat to a steam generator, which then drives the electricity-generating turbines.

Advantages: Compact design, high efficiency, and widespread global adoption.

Disadvantages: Requires enriched uranium and high-pressure systems.

Pressurized Water Reactor (PWR)

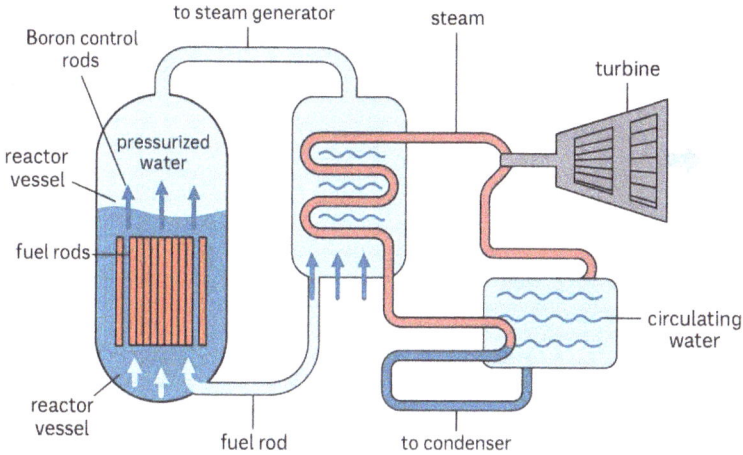

Table 5: Comparison Between Graphite Reactor and PWR

Characteristic	Graphite + Light Water (Obninsk)	PWR – Pressurized Water Reactor (Shippingport)
Moderator	Graphite	Light Water
Fuel	Natural uranium	Enriched uranium
System Pressure	Low	High

Source: Own elaboration based on the data presented in the Summary Table at the end of this chapter

Growth in the Number of Reactors and Installed Nuclear Capacity

During the 1960s and 1970s, there was significant growth in the number of nuclear reactors and the installed nuclear capacity worldwide.

Below, we present illustrative charts of this growth:

Chart 13: Evolution of the Number of Nuclear Reactors (1950–1970)

Growth in the Number of Nuclear Reactors Worldwide (1950-1970)

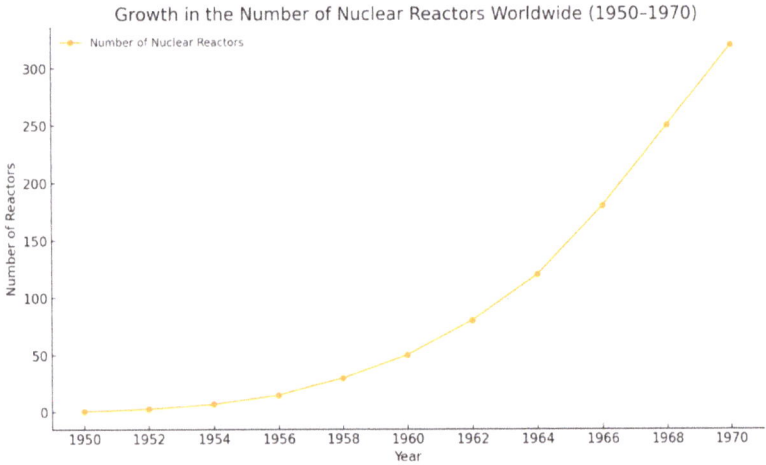

Source: Own elaboration based on the data presented in the Summary Table at the end of this chapter

Chart 14: Evolution of Installed Nuclear Capacity (1950–1970)

Growth of Installed Nuclear Capacity Worldwide (1950-1970)

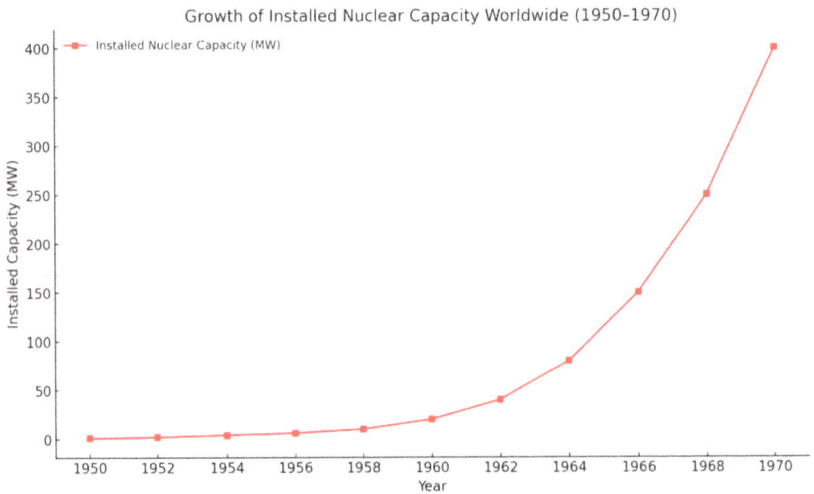

Source: Own elaboration based on the data presented in the Summary Table at the end of this chapter

Note: The above charts are approximate representations based on available historical data.

The Birth of the International Atomic Energy Agency (IAEA)

In response to growing concerns, the International Atomic Energy Agency (IAEA) was established in 1957 as an autonomous organization under the auspices of the United Nations, with its headquarters located in Vienna, Austria. The IAEA's main objectives include:

- **Promoting the Peaceful Use of Nuclear Energy**: Encourage and assist countries in developing and applying nuclear technologies for peaceful purposes, such as medicine, agriculture, and electricity generation.

- **Ensuring Safety and Protection**: Establish safety standards to protect people and the environment from the harmful effects of radiation.

- **Preventing Nuclear Proliferation**: Implement safeguards to ensure that nuclear materials and technologies are not diverted to the manufacture of nuclear weapons.

Global Importance of the IAEA

The IAEA plays a crucial role in regulating and monitoring nuclear activities worldwide. Its functions include:

- Regular Inspections: Verify nuclear facilities to ensure compliance with international agreements.

- Technical Assistance: Provide support, training, and capacity building to developing countries for the safe and effective use of nuclear technology.

- Forum for Cooperation: Serves as a platform for information exchange and collaboration among nations on nuclear issues.

The creation of the IAEA marked a milestone in global nuclear governance, striking a balance between promoting the benefits of nuclear technology and mitigating the risks associated with its misuse.

IAEA Interventions in Critical Situations

The IAEA has played a fundamental role in overseeing the use of nuclear materials, ensuring they are used exclusively for peaceful purposes. Over the years, the agency has intervened in several critical situations, particularly in countries suspected of developing nuclear weapons secretly. Notable interventions include:

Inspections in Iran

The case of Iran is one of the most prominent in the IAEA's work, with continuous inspections due to suspicions that the country was pursuing nuclear weapons under the guise of a civilian program.

- Discovery of Undeclared Facilities (2002): In 2002, an Iranian dissident group revealed the existence of undeclared nuclear facilities in Natanz and Arak, prompting rigorous IAEA inspections.

- 2015 Deal – Joint Comprehensive Plan of Action (JCPOA): The IAEA played a key role in this nuclear agreement between Iran and the world powers (the US, Russia, China, France, the UK, and Germany), under which Iran agreed to reduce its nuclear activities in exchange for the lifting of sanctions.

- Post-US Withdrawal Crisis (2018): In 2018, Donald Trump withdrew the US from the deal, and Iran began enriching uranium beyond permitted levels. Since then, the IAEA has carried out ongoing inspections to assess Iran's proximity to developing nuclear weapons.

Inspections in Iraq (1981 and 1991)

- Operation Opera (1981): Israel destroyed Iraq's Osirak nuclear reactor, claiming Saddam Hussein intended to develop nuclear weapons. The IAEA, which had been inspecting the program, was criticized for failing to detect the alleged military use.

- Post-Gulf War (1991): IAEA inspections revealed that Iraq had an advanced clandestine nuclear weapons program. As a result, Iraq was banned from conducting any nuclear activity until 2003.

Inspections in North Korea

North Korea signed the Nuclear Non-Proliferation Treaty (NPT) but has been under IAEA scrutiny since the 1990s.

- Expulsion of Inspectors (2002–2003): In 2002, the IAEA discovered North Korea was secretly enriching uranium. The country expelled inspectors and withdrew from the NPT in 2003.

- Nuclear Tests (2006–present): Since then, North Korea has conducted several nuclear tests, defying the international community. The IAEA remains barred from the country but monitors its activities via satellite and external intelligence.

Cases of Libya and Syria

- Libya (2003–2004): The Gaddafi regime admitted to a secret nuclear program with Pakistani technology. In 2004, following international negotiations, Libya allowed the dismantling of its program under IAEA supervision.

- Syria (2007): The IAEA investigated the bombing of a suspected nuclear reactor in Syria by Israel. The Syrian regime denied inspectors access, hindering verification efforts.

The IAEA continues to play an essential role in preventing nuclear proliferation, conducting technical inspections, and promoting agreements to ensure that nuclear energy is used exclusively for peaceful purposes.

In conclusion, the global expansion of nuclear energy between the 1950s and 1970s was marked by significant technological advancements and the need for international oversight structures. The introduction of different reactor types, such as graphite-moderated and PWRs, allowed for technological diversification. At the same time, the creation of the IAEA ensured that the growth of this energy source occurred in a safe and peaceful manner, establishing standards and promoting international cooperation.

Table 6: Core Functions of the IAEA

Function	Description
Promotion of Peaceful Use	Supports the use of nuclear energy for civilian purposes such as electricity and medicine.
Safety and Protection	Establishes international standards for nuclear safety and radiation protection.
Safeguards and Inspections	Prevents nuclear proliferation through site inspections and audits.

Source: Own elaboration based on the data presented in the Summary
Table at the end of this chapter

Global Expansion and Early Crises (1970–1990)

The 1970s marked a period of significant growth in nuclear energy, with many countries investing in this technology as an alternative to fossil fuels. However, as reactors proliferated across the globe, security challenges began to emerge.

Map showing the proliferation of nuclear power plants up to 1990, with circles representing each country's nuclear capacity installed. The larger the circle, the greater the nuclear capacity at that time.

Source: Own elaboration based on the data presented in the Summary Table at the end of this chapter

Two major accidents—Three Mile Island (1979) in the United States and Chernobyl (1986) in the Soviet Union—raised serious questions about the risks of nuclear energy. In addition, anti-nuclear movements gained momentum, leading to a slowdown in the sector in the West, while the Soviet Union and some Asian countries continued their nuclear expansion.

Growth of the Nuclear Sector in the U.S., Europe, the USSR, and Asia

In the 1970s, nuclear energy was widely promoted as the energy solution of the future due to several factors:

- Oil Crises (1973 and 1979): The surge in oil prices encouraged countries to seek alternative energy sources.

- Technological Advancements: Improvements in reactor safety and efficiency made nuclear energy more attractive.

- Industrial Expansion: Rising electricity demand spurred investments in nuclear power plants.

During this period, several countries significantly expanded their nuclear infrastructure:

United States: Became the world leader in nuclear energy, with more than 100 reactors in operation by the late 1980s.

Western Europe: Countries such as France, the United Kingdom, and Germany heavily invested in reactor construction. France, in particular, developed one of the most robust nuclear programs in the world.

Soviet Union: Continued expanding its nuclear program, building large-scale power plants, and developing the controversial RBMK reactor used in Chernobyl.

Asia: Japan and South Korea launched ambitious nuclear programs, viewing nuclear energy as a reliable and safe alternative to reduce dependence on imported oil.

Chart 15: Growth in the Number of Nuclear Reactors by Region (1960–2025)

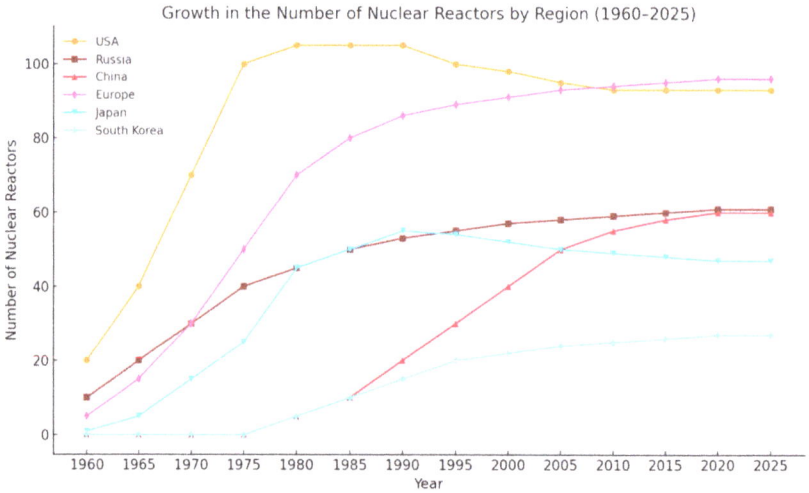

Source: Own elaboration based on the data presented in the Summary Table at the end of this chapter

This growth, however, was not without challenges. As the number of reactors increased, so did concerns about safety and the management of radioactive waste.

Early Nuclear Accidents and the Shift in Public Opinion

The optimism surrounding nuclear energy was severely shaken by two accidents that changed public perception of the risks associated with the technology.

Three Mile Island (USA, 1979) – The First Major Nuclear Crisis

The first major scare occurred on March 28, 1979, at the Three Mile Island nuclear power plant in Pennsylvania, USA.

What happened?

- A failure in the reactor's cooling pump led to overheating of the core.

- A human error worsened the situation, allowing part of the nuclear fuel to melt.

- Small amounts of radioactive gases were released into the environment.

Consequences:

- The accident caused no direct fatalities, but it generated widespread panic and a major crisis of confidence in the nuclear sector.

- The U.S. government halted the construction of new nuclear plants for several years.

- The incident highlighted the need for better safety protocols and operator training.

Three Mile Island was a turning point for nuclear energy in the United States, leading to stricter regulations and a decline in investment in new plants across the West.

Chernobyl (USSR, 1986) – The Disaster that Changed Global Nuclear Policy

If Three Mile Island was a warning, Chernobyl in 1986 was a catastrophe. To this day, it is considered the worst nuclear disaster in history.

What happened?

On the night of April 25–26, 1986, a team of engineers was conducting safety tests on Reactor 4 at the Chernobyl Nuclear Power Plant in Ukraine.

- Serious procedural errors led to an uncontrolled power surge in the RBMK reactor.

- The temperature reached extreme levels, and the reactor exploded, releasing a massive radioactive cloud.

Consequences:

- Immediate death of 31 workers and firefighters due to extreme radiation exposure.

- Evacuation of approximately 116,000 people from the region around the city of Pripyat.

- Radioactive contamination spread across Europe, affecting the health of millions.

- The Soviet Union attempted to conceal the disaster, but it quickly became evident that RBMK reactor safety was severely lacking.

Global Impact:

- Strengthened safety standards for nuclear reactors.

- Accelerated the decline of nuclear energy in Western Europe.

- Increased public resistance to the use of nuclear energy.

Chernobyl transformed global nuclear policy and exposed the dangers of a lack of transparency and inadequate safety in the nuclear sector.

Anti-Nuclear Movements and the Slowdown of Projects in the West

The accidents at Three Mile Island and Chernobyl fueled the rise of anti-nuclear movements, particularly in Europe and the United States.

Main arguments of anti-nuclear movements:

- Risk of new catastrophic accidents.

- Issues related to the storage of radioactive waste.

- High construction and maintenance costs of nuclear plants.

- Environmental and social impacts of power plants and uranium mining.

As a result of this pressure:

- Several countries, including Germany, Sweden, and Italy, canceled plans for new nuclear power plants.

- The growth of nuclear energy slowed in the West while Asia and the Soviet Union continued to expand their nuclear programs.

Nuclear energy entered a period of stagnation in the 1990s, particularly in the West, reflecting the fear and uncertainty sparked by these accidents.

Chart 16: Comparison Between the West and Russia + Asia in the Use of Nuclear Reactors

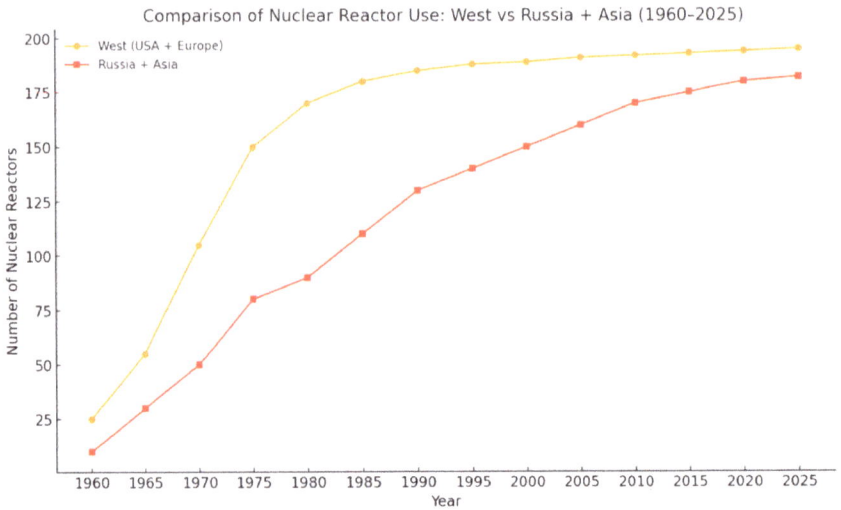

Comparison of Nuclear Reactor Use: West vs Russia + Asia (1960-2025)

Source: Own elaboration based on the data presented in the Summary Table at the end of this chapter

We can conclude that between 1970 and 1990, nuclear energy experienced both its greatest expansion and its first major challenges.

The nuclear sector grew rapidly, especially in the United States, Europe, the Soviet Union, and Asia.

The accidents at Three Mile Island and Chernobyl had a massive impact on public opinion and introduced new safety concerns.

Anti-nuclear movements gained strength, leading to a slowdown in nuclear energy in the West, while the USSR and some Asian countries continued to invest in the technology.

This period marked the end of absolute optimism about nuclear energy and the beginning of an era of tighter regulation and greater public scrutiny of the sector.

In the upcoming chapters, we will examine how the nuclear sector transformed itself in the aftermath of Chernobyl and how the pursuit of safety and technological innovation influenced the development of nuclear energy in the subsequent decades.

The Post-Chernobyl Period and the Nuclear Renaissance (1990–2010)

The 1990s marked a significant slowdown in the expansion of nuclear energy, especially in Western countries. The Chernobyl accident in 1986 left a deep imprint on public and political perceptions regarding the safety of nuclear power plants. Many nations began imposing stricter regulations, which increased costs and made new projects more difficult.

In the United States and Europe, the growth of nuclear energy stagnated. Some countries, such as Germany and Sweden, announced plans to gradually phase out their reliance on nuclear energy, opting instead for sources considered safer and more sustainable. Natural gas and renewable energies, such as wind and solar, have gained a stronger presence in the global energy mix.

In addition to public pressure, financial costs also posed a significant challenge. The construction of new plants became excessively expensive due to stricter safety requirements and extended development timelines. As a result, many companies in the nuclear sector faced financial difficulties, and several projects were canceled.

Modernizing Reactors for Greater Safety

With the turn of the millennium, concerns about climate change and the search for low-carbon energy sources reignited interest in nuclear energy. Countries like France and the United Kingdom began reassessing their energy policies, while China and Russia started investing heavily in the construction of new plants.

Nuclear technology also evolved during this period, with the development of so-called third-generation reactors designed to be safer and more efficient. Two of the most critical reactors from this era are the EPR (European Pressurized Reactor) and the AP1000, both of which incorporate significant advances in safety and efficiency.

The EPR Reactor (European Pressurized Reactor)

EPR is one of the most advanced pressurized water reactors (PWRs) ever developed. Designed by the French company Framatome and the German company Siemens, the EPR has a generation capacity of approximately 1,600 MWe, making it one of the most powerful reactors in the world. Its design incorporates multiple redundant safety systems and improved energy efficiency.

Key features include:

- **Enhanced safety systems**: include a double containment core and passive cooling systems that minimize the risk of core meltdown.

- **Higher thermal efficiency** enables better fuel utilization and reduces radioactive waste generation.

- **Extended operational life**: designed to operate for up to 60 years, with improved structural materials and components.

Currently, EPR reactors are in operation in countries such as France, China, and Finland, with more units planned for the future.

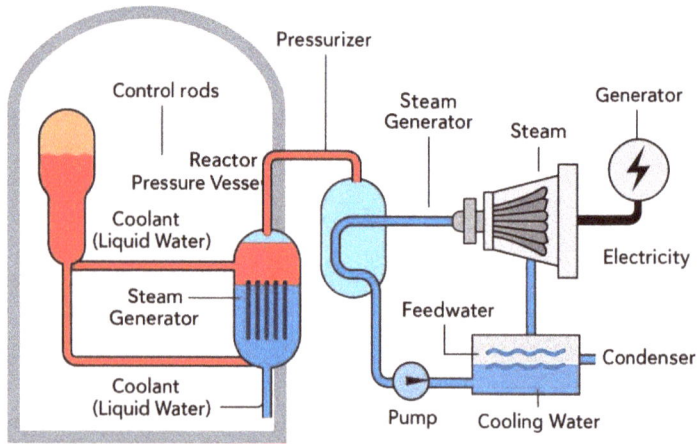

Diagram of the European Pressurized Reactor (EPR)

The EPR is a third-generation pressurized water reactor developed to offer enhanced safety and efficiency. Its main components include:

- Reactor Vessel: contains the core where nuclear fission occurs.

- Steam Generators: transfer heat from the primary to the secondary circuit, producing steam that drives the turbines.

- Pressurizer: maintains pressure in the primary circuit, preventing the formation of steam bubbles.

- Safety Systems: include multiple containment barriers and redundant systems to ensure reactor integrity in emergency situations.

The AP1000 Reactor

The AP1000, developed by the American company Westinghouse, is also a pressurized water reactor but with an innovative design focused on simplicity and passive safety. Its generation capacity is approximately 1,100 MWe.

Key features include:

- Passive Safety: uses safety systems that operate without electrical power or human intervention, ensuring continuous reactor cooling in the event of a severe failure.

- Fewer Components: reduced complexity in construction and maintenance, which lowers operational costs.

- Faster Construction: designed to be modular, enabling quicker construction and installation compared to earlier reactors.

The AP1000 has been primarily adopted in China, where several units are in operation. In the United States, its deployment has faced delays and financial challenges, but it remains a benchmark in nuclear sector innovation.

AP1000 Reactor

Passive Cooling Water Tank

Steam Generators

Steam

Passive Related Vent Valves

Turbine

Reactor Vessel

Pressurizer

Passive Residual Heat Removal Heat Exchanger

Steam

Feedwater Pump

Passive Related Heat Removal Heat- Exchanger

Condenser

Primary circuit
Secondary circuit
Passive safety systems

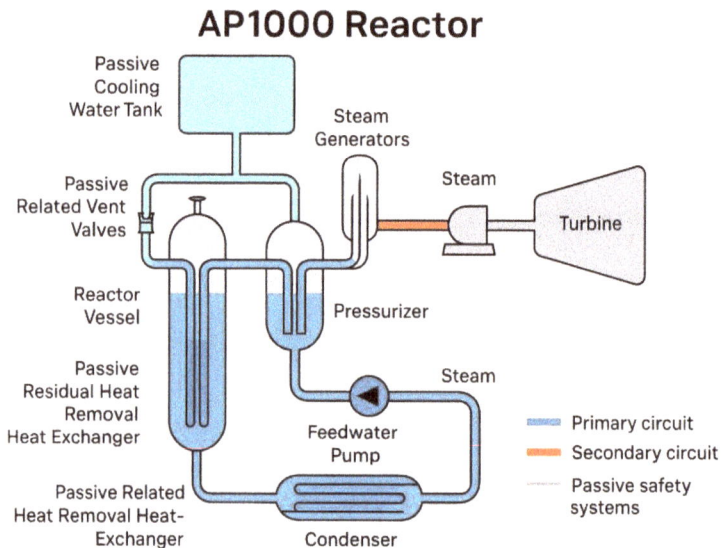

The AP1000, developed by Westinghouse, is a pressurized water reactor that incorporates passive safety systems. Its main components include:

- Reactor Vessel: houses the core and is designed to facilitate natural coolant circulation.

- Steam Generators: essential for heat transfer, with a simplified design to increase efficiency.

- Passive Safety Systems: rely on natural forces such as gravity and convection to cool the reactor without human intervention or external power for up to 72 hours.

New Countries Adopting Nuclear Technology and the Rise of SMRs

Despite Western hesitation, Asia emerged as the new epicenter of nuclear energy. China and India have heavily invested in nuclear technology, with dozens of reactors under construction to meet their growing electricity demands. Additionally, these countries began developing more advanced reactors, including Small Modular Reactors (SMRs) and nuclear fusion research.

Small Modular Reactors (SMRs)

Small Modular Reactors (SMRs) represent a recent innovation in nuclear technology, designed to offer flexibility, enhanced safety, and lower implementation costs compared to conventional reactors. They are smaller in scale, typically generating between 50 and 300 MWe, and offer significant advantages:

- Modular construction: enables factory fabrication and transportation to the operational site, reducing construction time and costs.

- Enhanced safety: Many SMRs use passive safety systems, minimizing accident risks.

- Versatility: can be deployed in remote locations or used to supply energy to smaller grids and specific industries.

- Diverse applications: in addition to electricity production, they can be used for water desalination, hydrogen production, and industrial heat.

Major SMR projects include:

- NuScale (USA): one of the most advanced projects, with regulatory approval and deployment plans underway.

- BWRX-300 (GE Hitachi, USA/Canada): based on boiling water reactor technology, promises lower costs.

- SMR-160 (Holtec, USA): focused on safety and ease of deployment.

- CAREM Reactor (Argentina): one of the first SMRs developed in Latin America.

Schematic Illustration

To better understand the functioning of SMRs, a schematic diagram is presented below, highlighting their main features, including modular system design, reactor configuration, and passive safety systems.

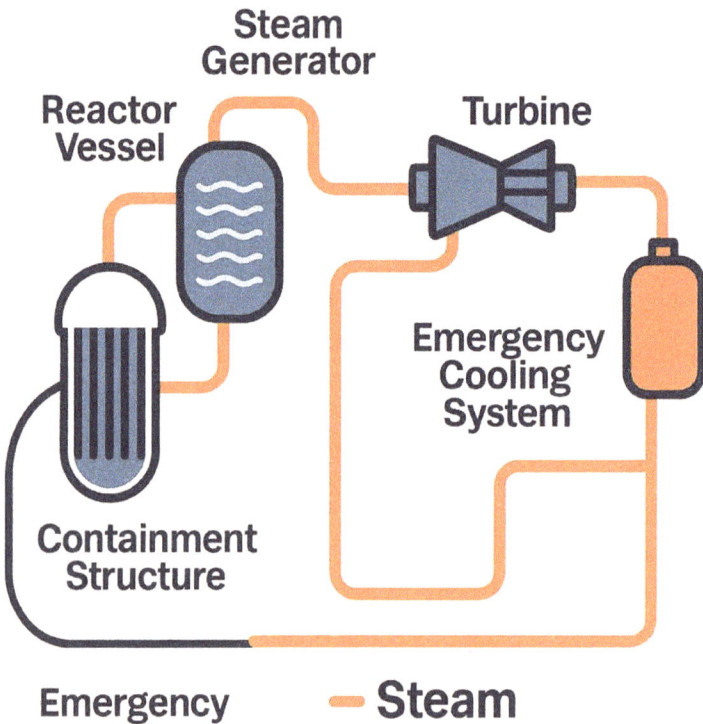

Steam Generator

Reactor Vessel

Turbine

Emergency Cooling System

Containment Structure

Emergency ▬ **Steam**

Main Components of a Small Modular Reactor (SMR):

- **Reactor Core**: Contains the nuclear fuel where fission occurs.

- **Primary Cooling System**: Circulates the coolant (water, gas, or molten salt) to remove heat generated in the core.

- **Steam Generator**: Transfers heat from the primary system to the secondary system, producing steam that drives electric turbines.

- **Passive Safety Systems**: Use natural physical principles, such as convection and gravity, to ensure safety without human intervention or external power.

France's Rise as a Global Leader in Nuclear Energy

France has established itself as one of the world's leading nuclear powers, with a trajectory marked by continuous investment and an energy policy centered on nuclear power.

Historical Evolution of the Number of Reactors in France

France's nuclear expansion began in the 1970s with the construction of multiple nuclear power plants aimed at reducing dependence on imported fossil fuels. Today, France operates 56 nuclear reactors, which supply around 70% of the country's electricity.

Future Plans and Projected Expansion

In 2022, the French government announced an ambitious nuclear expansion plan:

- Construction of six new EPR2 reactors, with the first expected to enter operation by 2035.

- Development of the NUWARD project, a small modular reactor (SMR) with a capacity of 340 MWe, slated for deployment by 2035.

Below, we present a chart illustrating the evolution of the number of nuclear reactors in France since the 1970s, along with projections for the coming years.

Chart 17: Evolution of the Number of Nuclear Reactors in France

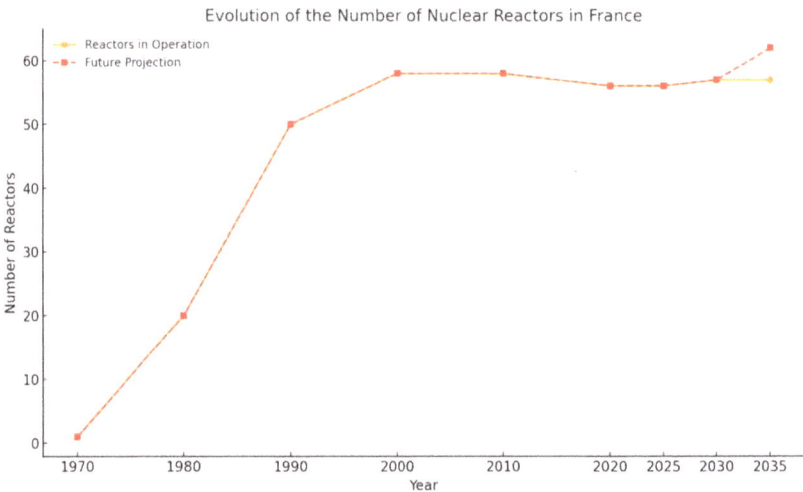

Evolution of the Number of Nuclear Reactors in France

Source: Own elaboration based on the data presented in the Summary Table at the end of this chapter

Russia's Role in Exporting Nuclear Technology

Russia has played a crucial role in the global expansion of nuclear energy by providing financing and technology to various countries. Rosatom has established strategic partnerships for the construction of reactors in regions looking to diversify their energy sources, such as the Middle East, Africa, and Southeast Asia.

Russia's strategy involved offering long-term financing and nuclear fuel supply contracts, ensuring a sustainable model for countries interested in adopting nuclear energy. This approach strengthened Russia's position as one of the leading global powers in the nuclear sector.

Analysis of Key Exporters:

- **Russia**: Rosatom, Russia's state-owned nuclear energy corporation, leads the global market, accounting for 76% of global nuclear technology exports. As of December 2030, the company was involved in constructing 35 nuclear power units in 12 different countries.

- **South Korea**: South Korea has emerged as a significant exporter of nuclear reactors, with agreements signed with the United Arab Emirates, Jordan, and Argentina. In 2024, Korea Hydro & Nuclear Power (KHNP) won a $17 billion project in the Czech Republic, surpassing competitors from the U.S. and France.

- **China**: China has heavily invested in nuclear energy both domestically and internationally. The country is involved in reactor construction in various nations, seeking to expand its influence in the global nuclear technology market.

- **France**: France, through companies such as Orano and EDF, maintains a significant presence in the nuclear technology market. In 2024, Orano announced the expansion of its uranium enrichment capacity in both France and the United States, aiming to reduce its dependence on Russian suppliers.

This chart and analysis highlight the concentration of the nuclear technology export market in a few countries, with Russia maintaining a dominant position, followed by key players such as South Korea, China, and France.

Chart 18: Leading Nuclear Technology Exporting Countries

Top Nuclear Technology Exporting Countries

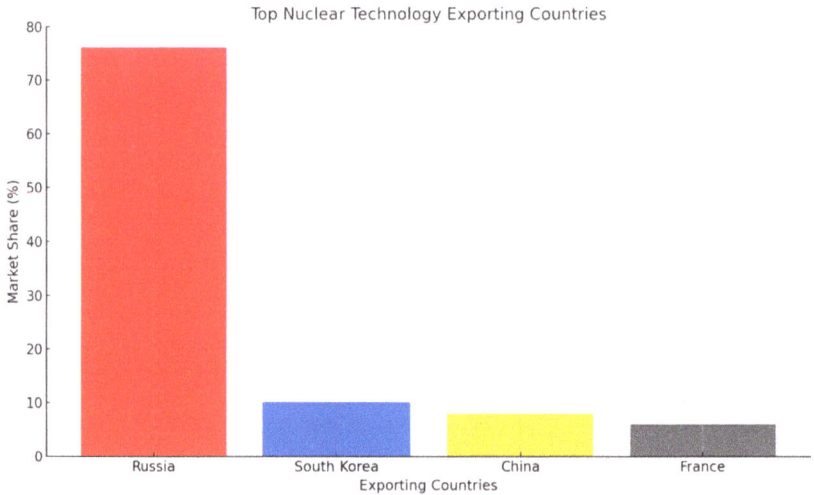

Source: Own elaboration based on the data presented in the Summary Table at the end of this chapter

Between 1990 and 2010, nuclear energy faced significant challenges, including disasters and disinvestment, as well as a strategic resurgence. While the West hesitated, Asia and Russia drove the sector forward, reinforcing its relevance in the global energy landscape. Safety, costs, and public acceptance remained central issues, but nuclear power continued to be a viable option for low-carbon energy production in the future.

Moreover, from an economic perspective, although the initial construction of nuclear power plants involves high upfront costs, their operation and maintenance are significantly more economical compared to other energy sources. The energy potential of nuclear power is vast and virtually inexhaustible, ensuring a continuous and reliable electricity supply in the long term. This makes nuclear energy one of the most cost-effective and sustainable options for global energy production.

The Fukushima Effect and the Future of Nuclear Energy (2011–Present)

On March 11, 2011, a magnitude 9.0 earthquake, followed by a devastating tsunami, struck Japan, causing one of the worst nuclear disasters in history: the accident at the Fukushima Daiichi nuclear power plant. The tsunami, with waves exceeding 14 meters, flooded the plant, disabling its cooling systems and leading to the partial meltdown of three reactor cores.

The consequences of the accident were severe:

- **Release of radioactive material**: large amounts of radiation were released into the atmosphere and ocean.

- **Mass evacuation**: more than 160,000 people were displaced due to the risk of contamination.

- **Global impact on public opinion**: reignited concerns about nuclear safety and prompted governments to reassess their nuclear programs.

- **High costs**: cleanup, compensation, and plant decommissioning costs were estimated in hundreds of billions of dollars.

The Fukushima accident led to new global regulations aimed at improving the safety of nuclear power plants, including stricter requirements for cooling systems and protection against natural disasters.

Countries That Reduced or Abandoned Their Nuclear Programs

The Fukushima accident prompted several countries to reconsider their use of nuclear energy. Among those that significantly reduced or completely abandoned their nuclear programs are:

- **Germany** announced a complete nuclear phase-out plan, progressively shutting down its reactors. In April 2023, the country decommissioned its last nuclear reactors.

- **Italy** had already halted its nuclear program following a 1987 referendum but completely ruled out any revival after Fukushima.

- **Switzerland and Belgium** decided not to build new reactors and set plans to reduce their dependence on nuclear energy.

The decisions made by these countries were driven by a combination of factors, including public pressure, perceived risks, and the growing feasibility of renewable energy sources.

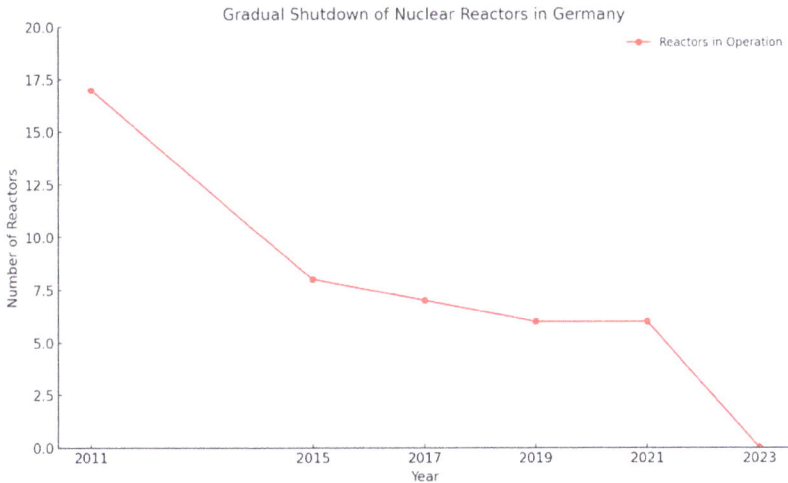

Source: Own elaboration based on the data presented in the Summary
Table at the end of this chapter

In 1987, Italy held three national referendums related to nuclear energy, resulting in a significant decision on the future of the sector in the country. These referendums took place on November 8, 1987, with a voter turnout of 65.1%.

Referendum Topics and Results:

- Location of Nuclear Power Plants:

 Question: Should the state's power to impose the construction of nuclear plants in municipalities that did not agree be abolished?

 Result: 80.6% voted "Yes" to abolish this state power.

- Financial Incentives for Municipalities:

Question: Abolish financial incentives offered to municipalities that accept the construction of nuclear or coal plants.

Result: 79.7% voted "Yes" to eliminate these incentives.

- ENEL's Participation in International Nuclear Projects:

Question: Prohibit ENEL (the National Electricity Company) from participating in the construction and management of nuclear plants abroad.

Result: 71.9% voted "Yes" to prohibit this participation.

Referendum Consequences:

Although the questions were technical and did not explicitly ban nuclear energy, the results reflected public opposition to nuclear power following the 1986 Chernobyl disaster. As a result, the Italian government began shutting down existing nuclear plants:

- Construction of the nearly completed Montalto di Castro nuclear plant was halted.

- The Caorso and Enrico Fermi plants were decommissioned in 1990.

- The Latina plant had already been shut down in December 1987.

These actions marked the end of nuclear energy production in Italy, a decision that remains in effect to this day.

Current Status of Switzerland and Belgium:

Switzerland:

- Number of operating reactors: 4

- Number of nuclear power plants: 3

Belgium:

- Number of operating reactors: 5

- Number of nuclear power plants: 2

In terms of electricity generation, nuclear reactors accounted for 36.4% of Switzerland's total electricity production in 2022 and 46.4% in Belgium during the same year.

It is worth noting that both countries have plans to gradually reduce their reliance on nuclear energy. In Switzerland, a 2017 referendum approved a ban on building new nuclear power plants, aiming for a complete phase-out of nuclear power by 2050. In Belgium, a 2003 law mandated a gradual nuclear phase-out by 2025; however, due to factors such as the war in Ukraine and rising gas prices, the government decided to extend the operation of two of the country's seven nuclear reactors until 2035.

Countries That Continued Investing in Nuclear Energy

Despite the impact of Fukushima, some countries reaffirmed or even expanded their nuclear programs, recognizing its role in energy security and decarbonization:

- **France**: Maintained its firm reliance on nuclear power, which accounts for about 70% of the country's electricity. The French government announced plans to build new EPR2 reactors and expand research in nuclear fusion.

- **China**: Intensified its nuclear program, with dozens of new reactors under construction. The country views nuclear energy as a crucial component of its clean energy strategy.

- **Russia**: Continued investing in the construction and export of nuclear reactors, including floating reactors to supply power to remote regions.

- **United States**: Although some plants were decommissioned, the country approved new projects, including next-generation reactors and investments in small modular reactors (SMRs).

Country Analysis:

- **United States**: In 2010, there were 104 operating nuclear reactors. By 2023, this number had decreased to 93, reflecting a trend of decommissioning older units without a corresponding increase in the construction of new ones.

- **France**: Maintained a stable policy toward nuclear energy, with a slight reduction from 58 reactors in 2010 to 56 in 2023. Nuclear power remains the primary source of electricity, accounting for about 70% of total production.

- **China** Showed significant growth in the nuclear sector, increasing from 13 reactors in 2010 to 54 in 2023. This growth reflects the country's strategy to diversify its energy sources and reduce dependence on fossil fuels.

- **Russia**: Recorded moderate growth, increasing from 32 reactors in 2010 to 37 in 2023. In addition, Russia has become a major exporter of nuclear technology, with 26 units under construction, six domestically and 20 in seven other countries.

This chart illustrates how each country adjusted its nuclear energy investments after the Fukushima incident, with China showing particularly notable growth in the sector.

Chart 20: Investments in Nuclear Energy Before and After Fukushima

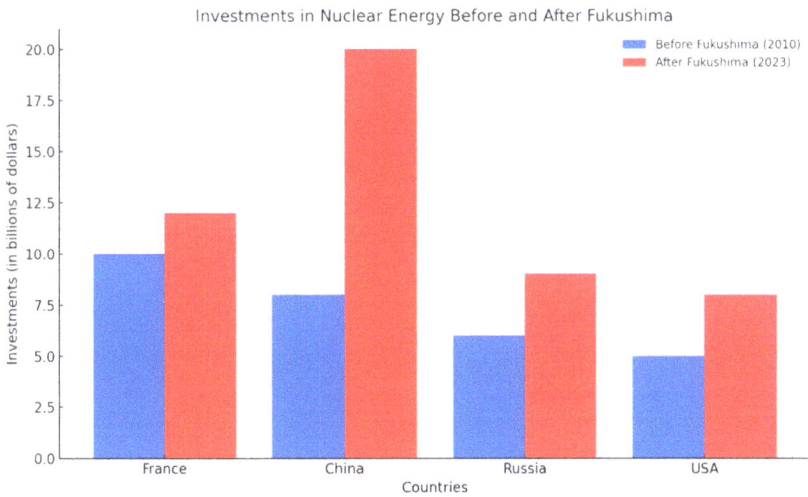

Source: Own elaboration based on the data presented in the Summary Table at the end of this chapter

Chart 21: Number of Nuclear Reactors Built After Fukushima

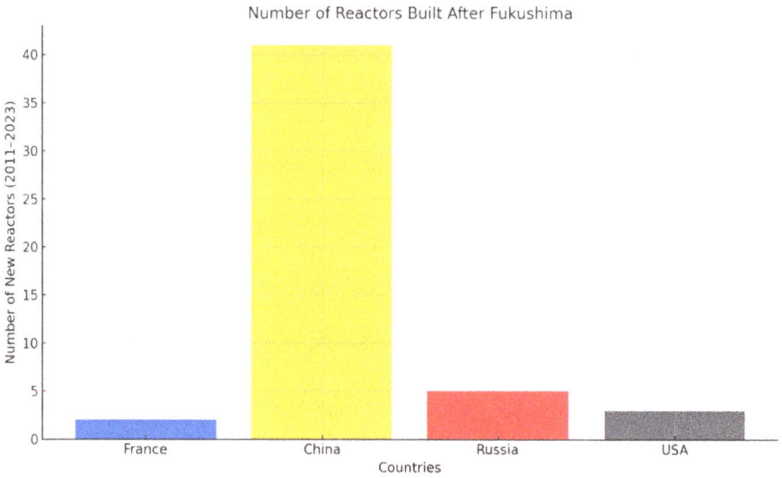

Number of Reactors Built After Fukushima

Source: Own elaboration based on the data presented in the Summary Table at the end of this chapter

Development of Small Modular Reactors (SMRs)

Currently, there are only a few Small Modular Reactors (SMRs) in operation worldwide. Russia and China are pioneers in this technology, each with distinct projects:

- **Russia**: Operates the "Akademik Lomonosov," a floating nuclear power plant equipped with two SMR units of 35 MW each, totaling 70 MW.

- **China**: In 2021, the HTR-PM reactor was connected to the power grid—a high-temperature gas-cooled modular reactor.

Currently, there are three SMRs in operation worldwide: two in Russia and one in China.

Chart 22: Number of SMR Reactors in Operation by Country

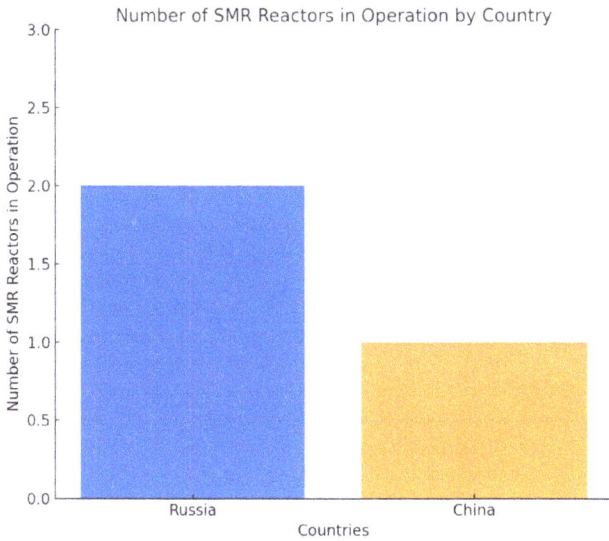

Number of SMR Reactors in Operation by Country

Source: Own elaboration based on the data presented in the Summary Table at the end of this chapter

Several countries are currently investing significantly in the research and development of Small Modular Reactors (SMRs). Below is a chart illustrating the main investors in this technology:

Chart 23: Projects by Country in SMR Technology

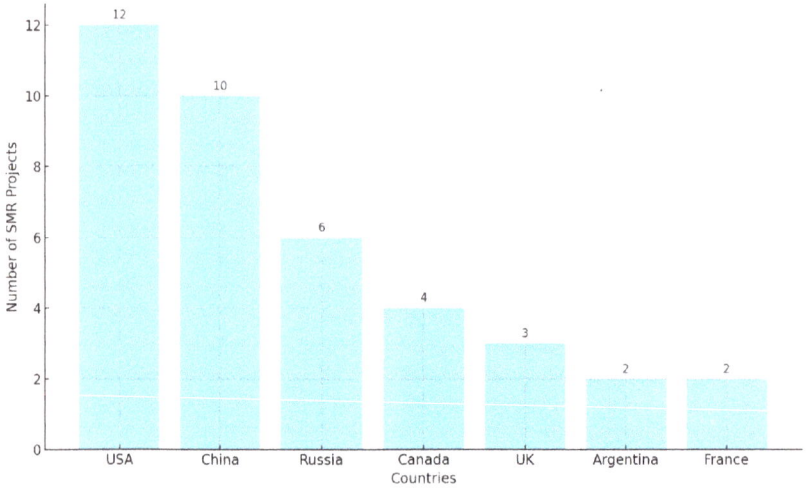

Source: Own elaboration based on the data presented in the Summary Table at the end of this chapter

Chart 24: Investment by Country in SMR Technology

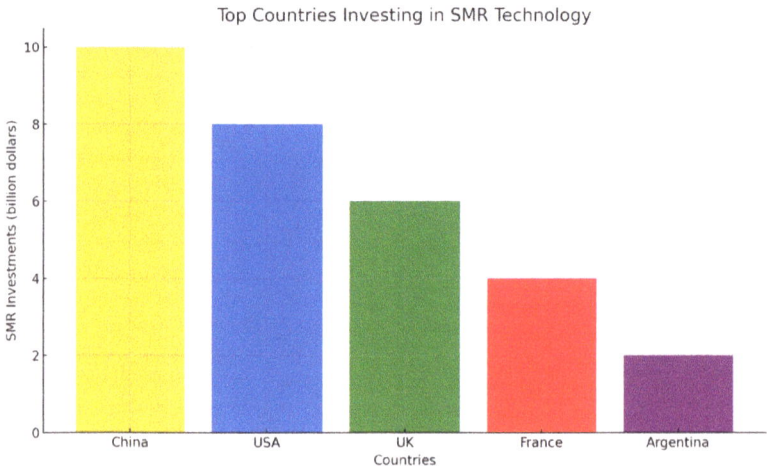

Source: Own elaboration based on the data presented in the Summary Table at the end of this chapter

Investment Analysis:

- **China:** Leads SMR investments, with approximately USD 10 billion allocated to the development and implementation of this technology.

- **United States**: Invested around USD 8 billion in SMR research and development, with companies like NuScale Power leading projects to build operational reactors by 2029.

- **United Kingdom**: Committed to around USD 6 billion in SMR construction, aiming to boost economic growth and provide clean and affordable energy.

- **France**: Allocated about USD 4 billion to SMR development, with EDF leading the Nuward project, which aims to build 340 MWe modular reactors.

- **Argentina**: Invested approximately USD 2 billion in the CAREM-25 project, a 25 MWe SMR prototype entirely designed and developed in the country.

In summary, we can conclude that the Fukushima accident was a turning point in the history of nuclear energy, prompting some nations to reduce or abandon their nuclear programs. However, other countries continued to invest in the technology, recognizing its role in enhancing energy security and reducing carbon emissions. The development of SMRs and the pursuit of nuclear fusion suggest that the nuclear sector may have a promising future, combining enhanced safety and new energy applications.

Conclusion of the Present Chapter

Nuclear energy has undergone several transformations since its discovery, evolving from experimental technology into one of the world's primary sources of electricity. From the first commercial reactors in the 1950s to modern advancements in safety and efficiency, nuclear technology has shown resilience and adaptability in the face of historical challenges such as accidents, political crises, and shifts in global energy policies.

With the growing need to reduce carbon emissions and ensure energy security, nuclear power is emerging as a key solution for the global energy transition. Its ability to generate continuous electricity, regardless of weather conditions, makes it a strong complement to intermittent renewable sources like solar and wind.

Investments in third-generation reactors, small modular reactors (SMRs), and nuclear fusion point to a future in which nuclear energy will play an even more significant role, delivering reliable and low-impact power.

The history of nuclear energy shows how past events—such as Chernobyl and Fukushima—have shaped public perception and government policies. However, these events have also driven significant improvements in safety and technology, enabling the development of more efficient and secure reactors.

Although the construction of nuclear power plants requires high initial investments, their operation and maintenance costs are significantly lower compared to other energy sources. Furthermore, the high energy density of uranium and the long lifespan of reactors ensure cheap and reliable power for decades.

The economic benefits of nuclear energy are reflected in reduced electricity costs for consumers and industries, making it a competitive alternative for ensuring a stable energy supply. With the advancement of technologies like SMRs, construction costs are expected to decrease, expanding access to nuclear energy in different regions of the world.

Nuclear energy remains a key element in the global energy matrix. Its economic impact, ability to provide clean and reliable electricity, and ongoing technological advancements reinforce its importance in transitioning to a more sustainable and secure energy future.

Table 7: Sources Consulted in Chapter 2

Books on the History of Science – Evolution of physics and nuclear discoveries.
Works of Marie Curie and Ernest Rutherford – Studies on radioactivity.
Reports from the World Nuclear Association – History of nuclear development.
Documents from the Manhattan Project – Development of the atomic bomb.
US National Archives – Einstein's letters to Roosevelt.
IAEA Publications – History of civilian nuclear energy.
Military History Books – World War II and nuclear energy.
Popular science works (e.g., Brian Cox, Richard Rhodes) – Nuclear Energy and Society.

BBC History and History Channel – Documentaries on the Manhattan Project and World War II.
Scientific American and Nature – Articles on the early days of nuclear fission.
Scientific journals and historical newspapers – Coverage of the nuclear age from 1930 to 1950.

Next Chapter: Nuclear Energy and Nuclear Weapons – Myths and Realities

In the next chapter, we will explore the relationship between nuclear energy and nuclear weapons, debunking misconceptions and analyzing the impact of this technology on geopolitics and global security.

Chapter 3 -
Nuclear Energy and Nuclear Weapons: Myths and Truths

Nuclear energy is often associated with nuclear weapons, leading to the mistaken perception that electricity generation through nuclear means is a direct pathway to building atomic bombs. This widespread view stems mainly from the historical impact of World War II and the Cold War arms race, which popularized the idea that any nuclear program could pose a threat.

However, the reality is quite different. Nuclear technology can be used peacefully, contributing to the generation of clean and efficient electricity while also driving advances in medicine, industry, and even space exploration. The development of nuclear weapons, on the other hand, requires highly specialized technical processes and uranium enrichment levels far beyond those used in commercial reactors.

Moreover, there is a clear distinction between countries that pursue peaceful nuclear programs—under strict oversight by the International Atomic Energy Agency (IAEA) and those that choose to develop nuclear weapons, usually in secrecy and under heavy international restrictions.

This chapter will clarify the fundamental differences between these two uses of nuclear technology and will debunk common misconceptions, explaining in a fact-based and objective manner why nuclear energy for electricity generation does not, in itself, represent a risk of atomic weapons proliferation.

Differences Between Nuclear Energy for Power Generation and Nuclear Weapons

Although both applications rely on similar nuclear physics principles, their purposes, processes, and materials involved are fundamentally different:

- **Purpose**: Civilian nuclear energy aims to generate electricity sustainably, whereas nuclear weapons are designed for large-scale destruction.

- **Materials Used**: The primary distinction between nuclear reactors and atomic bombs lies in the composition of nuclear fuel.

  - **Uranium in nuclear reactors**: Uranium used for electricity generation (U-235) is enriched to only 3–5%.

  - **Uranium for nuclear weapons**: Highly Enriched Uranium (HEU) used in bombs contains over 90% U-235.

  - **Plutonium**: Plutonium-239, used in nuclear weapons, is produced in specialized reactors and requires advanced reprocessing techniques.

Uranium Mining and Enrichment Process

Natural uranium is extracted from open-pit or underground mines and, after mining, undergoes a beneficiation process to remove impurities. Naturally occurring uranium contains only about 0.7% of U-235, the fissile isotope required for both energy generation and weapon construction.

Uranium enrichment is achieved through isotopic separation, with gas centrifugation being the most commonly used technique today. The process includes:

1. **Conversion to uranium hexafluoride (UF_6)**: Natural uranium is converted into gas to facilitate isotope separation.

2. **Centrifugation**: The UF_6 gas is introduced into high-speed centrifuges. Since U-238 is heavier than U-235, it concentrates near the outer edge, while the lighter U-235 collects toward the center.

3. **Repetition of the process**: The procedure is repeated in thousands of interconnected centrifuges (a centrifuge cascade) until the desired enrichment level is reached.

For nuclear reactors, uranium is enriched to 3–5%, while for nuclear weapons, enrichment exceeds 90%.

NUCLEAR FUEL CYCLE

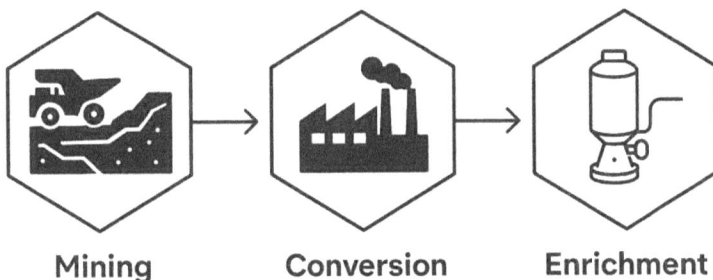

Mining   Conversion   Enrichment

Plutonium Production

Plutonium-239 is produced in nuclear reactors from uranium-238. During reactor operation, some U-238 atoms capture neutrons and transform into plutonium-239, a highly fissile isotope.

For military use, this plutonium must be extracted from irradiated fuel elements through a chemical process known as reprocessing, which involves:

1. Removal of the irradiated fuel from the reactor

2. Dissolution of the fuel in nitric acid

3. Chemical separation of plutonium-239 from the remaining fission products

International agencies strictly monitor this process, as the extraction of plutonium can indicate potential attempts to develop nuclear weapons.

Table 8: Methods of Obtaining Fissile Material

Method	Produced Material	Description
Uranium Enrichment (centrifugation)	U-235	Isotopic separation to increase the proportion of U-235.
Nuclear reactors + reprocessing	Pu-239	Plutonium obtained from U-238 irradiated in reactors.

Source: Own elaboration based on the data presented in the Summary Table at the end of this chapter

Reprocessing of Spent Nuclear Fuel

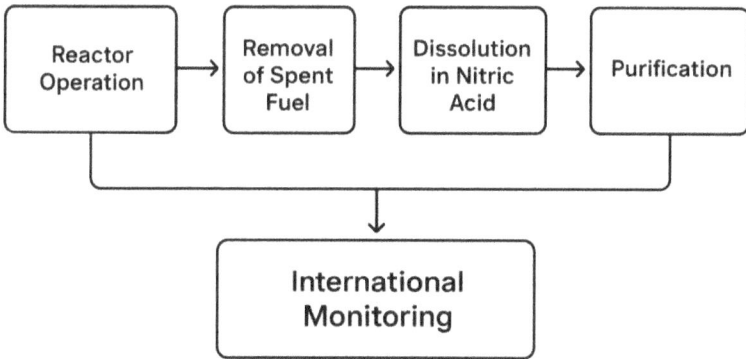

Here is the illustrative diagram of the plutonium production process, from reactor operation to purification and potential military use.

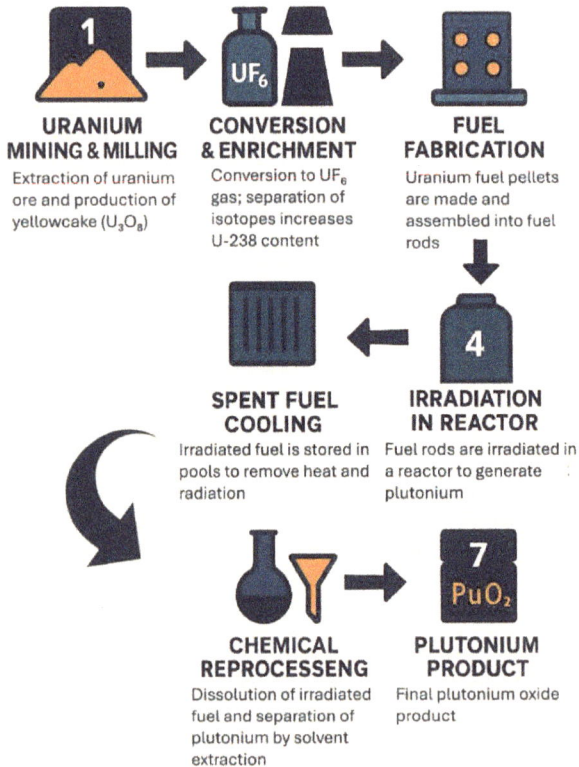

PLUTONIUM PRODUCTION CYCLE

URANIUM MINING & MILLING
Extraction of uranium ore and production of yellowcake (U_3O_8)

CONVERSION & ENRICHMENT
Conversion to UF_6 gas; separation of isotopes increases U-238 content

FUEL FABRICATION
Uranium fuel pellets are made and assembled into fuel rods

SPENT FUEL COOLING
Irradiated fuel is stored in pools to remove heat and radiation

IRRADIATION IN REACTOR
Fuel rods are irradiated in a reactor to generate plutonium

CHEMICAL REPROCESSENG
Dissolution of irradiated fuel and separation of plutonium by solvent extraction

PLUTONIUM PRODUCT
Final plutonium oxide product

The Myth of the Gateway to Nuclear Weapons

Many argue that any civilian nuclear program can serve as a cover for the development of nuclear weapons, but the reality is far more complex.

International Monitoring: Any country developing nuclear technology for peaceful purposes is subject to strict inspections by the International Atomic Energy Agency (IAEA), which ensures that nuclear materials are used solely for civilian applications.

Technical Challenges: Converting a civilian program into a weapons program requires specialized infrastructure, such as advanced centrifuges or reactors dedicated to plutonium production, as well as highly advanced technical expertise.

Reactor Types: Commercial light-water reactors (PWR and BWR) are not ideal for producing weapons-grade plutonium, as the nuclear fuel must be removed early to avoid contamination with undesirable isotopes.

International Treaties: The Nuclear Non-Proliferation Treaty (NPT) establishes limits and regulations to prevent the spread of nuclear weapons, requiring signatory countries to submit to audits and controls.

Table 9: Comparison of International Nuclear Treaties

Treaty	Objective	Signatories	Current Status
NPT	Prevent nuclear proliferation	191 countries	In force
CTBT	Ban nuclear testing	185 countries (not all ratified)	Not in force
TPNW	Ban nuclear weapons	92 signatory countries	In force since 2021

Source: Own elaboration based on the data presented in the Summary Table at the end of this chapter

The belief that any country developing nuclear energy will inevitably develop atomic weapons overlooks these factors and contributes to an alarmist and misinformed narrative. The existence of countries such as **Japan and Germany**, which possess advanced nuclear technology without developing weapons, reinforces the clear distinction between peaceful and military uses of nuclear energy.

In addition, countries such as **Canada, Brazil, Argentina, and South Korea** possess advanced nuclear technology and operate power reactors without pursuing nuclear weapons. All these nations are signatories of the Nuclear Non-Proliferation Treaty and maintain nuclear programs that the IAEA heavily monitors.

However, some nations are the subject of international suspicion due to potential military nuclear ambitions under the pretext of peaceful development. **Iran and Saudi Arabia**, for example, are frequently mentioned in proliferation debates due to their interest in uranium enrichment and lack of transparency in some regions of their nuclear programs.

Another well-known case is **North Korea**, which initially developed a nuclear program under the pretense of electricity generation but later withdrew from the Nuclear Non-Proliferation Treaty and tested nuclear devices, becoming a declared nuclear weapons state.

This demonstrates that nuclear energy can be used entirely peacefully and does not inherently lead to weapons development. The real differentiating factor lies in governance, international commitments, and active oversight by bodies such as the IAEA, which ensure that nuclear materials are used exclusively for civilian purposes.

Countries with Military Nuclear Capabilities

Countries with military nuclear capabilities are those that have developed, tested, and currently possess operational nuclear weapons arsenals. These nations can be divided into two main groups: those officially recognized under the Nuclear Non-Proliferation Treaty (NPT) and those that developed weapons outside of it.

NPT-Recognized Nuclear States:

The five countries officially recognized as nuclear powers by the NPT are:

- **United States**

- **Russia**

- **China**

- **France**

- **United Kingdom**

These nations have established and declared arsenals and are permanent members of the United Nations Security Council.

Nuclear-Armed States Outside the NPT

Other countries that developed nuclear weapons without official NPT recognition include:

- **India** – Conducted nuclear tests in 1974 and 1998, establishing itself as a nuclear power.

- **Pakistan** – Developed weapons in response to India, testing in 1998.

- **North Korea** – Withdrawn from the NPT and has conducted multiple tests since 2006.

- **Israel (presumed)** – Neither confirms nor denies possessing nuclear weapons but is widely believed to have a significant arsenal.

Table 10: Selected Geopolitical Cases

Country	NPT Status	Nuclear Program	Inspections/Allegations
Israel	Non-signatory	Undeclared but suspected	No IAEA inspections
Iran	Signatory	Civilian with suspicions	Regular inspections and sanctions
India	Non-signatory	Military and civilian	Civilian reactors under inspection
Pakistan	Non-signatory	Nuclear weapons	No IAEA inspections
North Korea	Withdrawn	Nuclear weapons	Limited access and declared tests

Source: Own elaboration based on the data presented in the Summary Table at the end of this chapter

Latent Nuclear Capacity

There are countries without declared arsenals but with the technical capacity to quickly develop nuclear weapons if they choose to do so. These include **Germany, Japan, South Korea, and Iran**. These nations have advanced nuclear programs and, theoretically, could produce weapons if they opted to.

Quantity and Destructive Power

Arsenals vary greatly, with the **U.S. and Russia** holding the largest stockpiles, consisting of thousands of active and stored warheads. China, France, and the UK maintain smaller but highly modernized arsenals.

The destructive impact of these weapons depends on the warhead type, with some bombs being hundreds of times more powerful than those dropped on Hiroshima and Nagasaki.

This unequal distribution of nuclear arms reflects global geopolitics and the ongoing challenges of non-proliferation. Arms control and disarmament agreements remain key topics in international diplomacy.

Table 11: Nuclear-Armed States and NPT Status

Country	NPT Signatory	Declared nuclear weapons
United States	Yes	Yes
Russia	Yes	Yes
China	Yes	Yes
France	Yes	Yes
United Kingdom	Yes	Yes
India	No	Yes
Pakistan	No	Yes
Israel	No	Not declared
North Korea	Withdrawn	Yes

Source: Own elaboration based on the data presented in the Summary Table at the end of this chapter

Relationship Between Civilian and Military Nuclear Programs

Not all countries with nuclear weapons also use this technology for peaceful purposes. There is an important distinction between those with **dual-use nuclear programs**, civilian and military—and those with only one of the two.

Countries with military capacity and robust civilian programs: The United States, Russia, France, and China possess both military arsenals and extensive programs for nuclear power generation, medical research, and peaceful technological development.

Countries with nuclear weapons but limited civilian infrastructure: Israel and North Korea have nuclear weapons but do not operate significant civilian nuclear energy programs.

Countries with advanced civilian programs but no nuclear weapons: Japan, Germany, and Canada are examples of nations with high nuclear technological capacity that have chosen not to develop military arsenals.

Chart 25: Comparison Between Investments in Peaceful vs. Military Nuclear Programs

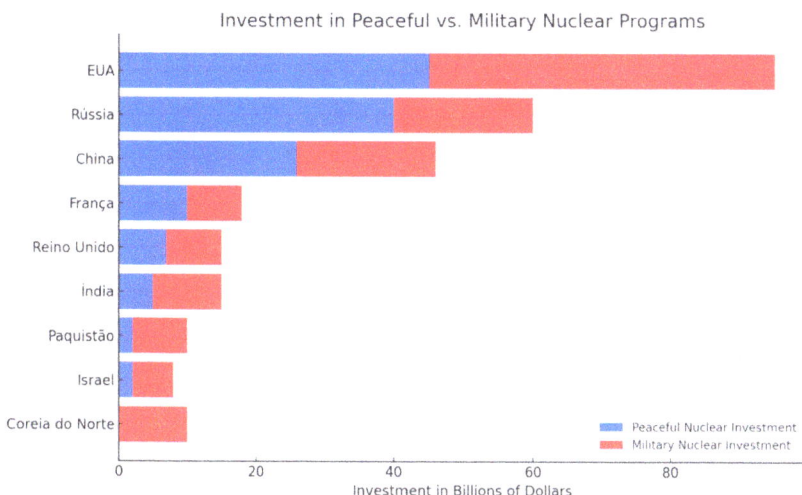

Source: Own elaboration based on the data presented in the Summary Table at the end of this chapter

Table 12: Dual-Use Nuclear Technologies

Technology	Civilian Use	Military/Potential Use
Uranium enrichment	Fuel for reactors	Raw material for nuclear bombs
Nuclear reactors	Electricity generation, medicine	Plutonium production
Lasers and accelerators	Scientific research	Development of advanced weapons

Source: Own elaboration based on the data presented in the Summary Table at the end of this chapter

Comparative chart of investment in peaceful vs. military nuclear programs by country. It illustrates the difference in budgets allocated to civilian nuclear energy and the development of nuclear weapons.

This scenario demonstrates that the possession of nuclear weapons is not directly linked to the peaceful use of nuclear energy, and vice versa. Many countries develop nuclear technology for peaceful purposes with no intention of militarization, while others maintain arsenals without investing in nuclear electricity generation.

Nuclear Warheads and Their Destructive Power

Nuclear warheads represent the most destructive type of weapon ever created by humanity. These explosive devices use nuclear reactions to release colossal amounts of energy in a very short period of time. Warheads can be mounted on various types of missiles, such as intercontinental ballistic missiles (ICBMs), submarine-launched ballistic missiles (SLBMs), and aerial bombs.

Structure and Function of Nuclear Warheads

Nuclear warheads can be divided into two main types:

Fission bombs (atomic bombs): These are based on the splitting of atomic nuclei of elements like uranium-235 or plutonium-239, releasing a large amount of energy. Historical examples include the bombs dropped on Hiroshima and Nagasaki.

Thermonuclear bombs (hydrogen bombs): These more advanced weapons use the fusion of hydrogen isotopes (deuterium and tritium), releasing significantly more energy than fission bombs.

A modern nuclear warhead contains:

1. **Primary charge (Nuclear Fission)**: A conventional explosive detonates a subcritical mass of fissile material, initiating a chain reaction.

2. **Secondary charge (Nuclear Fusion, in thermonuclear bombs)**: The energy from the initial explosion is used to compress and heat the fusion fuel, greatly increasing energy output.

3. **Detonation system**: Precision triggering mechanisms ensure detonation occurs only under authorized command.

4. **Shielding and protection**: Layers of durable materials provide safe transport and storage.

Destructive Power of Nuclear Warheads

The destructive power of a nuclear warhead is measured in kilotons (kt) or megatons (Mt) of TNT equivalent. For reference:

- **Hiroshima bomb (Little Boy)**: 15 kt – Destroyed an entire city and caused approximately 140,000 direct deaths.

- **Nagasaki bomb (Fat Man)**: 21 kt – Caused massive city destruction and 80,000 deaths.

- **Tsar Bomba (largest ever tested, Russia, 1961)**: 50 Mt – An explosion a thousand times more powerful than Hiroshima.

Modern bombs can be yield-adjustable, allowing the explosive power to be varied according to tactical needs.

Countries with the Largest Number of Nuclear Warheads

Nuclear arsenals vary significantly among global powers. Countries with the largest stockpiles of active and stored warheads include:

- **Russia** – Around 6,000 warheads.

- **United States** – Approximately 5,500 warheads.

- **China** – Around 500 warheads, rapidly expanding.

- **France** – About 290 warheads.

- **United Kingdom** – Approximately 225 warheads.

- **Pakistan** – Around 165 warheads.

- **India** – Estimated between 160 to 170 warheads.

- **Israel** – Though not officially confirmed, it is believed to possess 80 to 100 warheads.

- **North Korea** – Estimated between 40 to 50 warheads.

These figures represent both active and stockpiled warheads; however, each country has its own use doctrine, which influences its military strategies and defense policies.

Chart 26: Number of Nuclear Warheads by Country

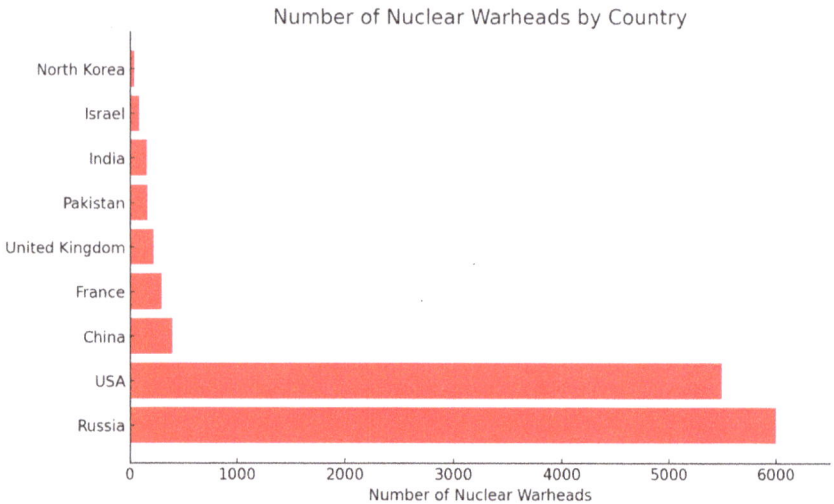

Number of Nuclear Warheads by Country

Source: Own elaboration based on the data presented in the Summary Table at the end of this chapter

Humanitarian and Environmental Consequences

The impact of a nuclear explosion extends far beyond the initial blast:

- **Shock wave**: Extreme pressure destroys buildings and infrastructure across several kilometers.

- **Intense heat**: Can incinerate entire cities and trigger massive fires.

- **Initial radiation and radioactive fallout**: Radiation can cause immediate deaths and long-term illnesses, as well as contamination of the environment for decades.

Thus, nuclear warheads represent an existential risk to humanity, which is why international agreements such as the Nuclear Non-Proliferation Treaty (NPT) and the Treaty on the Prohibition of Nuclear Weapons (TPNW) aim to limit their use and development.

To illustrate the section on missiles that carry nuclear warheads, I've selected some representative images of different launch systems used by various nations.

RS-24 Yars Intercontinental Ballistic Missile (ICBM), Russia: This missile is capable of carrying multiple nuclear warheads and has a range of up to 12,000 km.

Trident II Submarine-Launched Ballistic Missile (SLBM), USA: Used by the United States Navy, the Trident II is a nuclear-capable missile launched from submarines.

Hwasong-15 Ballistic Missile, North Korea: This intercontinental missile was tested by North Korea and is capable of carrying nuclear warheads.

These images illustrate the variety and sophistication of missile systems developed by different countries to carry nuclear warheads, highlighting the importance of understanding the capabilities and risks associated with these weapons.

Conclusion of the Present Chapter

The distinction between the peaceful and military uses of nuclear energy is one of the most crucial issues in modern geopolitics. While peaceful nuclear energy has served as a vital pillar for the development of many nations—ensuring stable electricity supply, advancements in medicine, and new technological applications—its military use represents one of humanity's greatest existential risks.

Responsible countries committed to global well-being utilize nuclear technology for peaceful purposes, abiding by international treaties such as the **Nuclear Non-Proliferation Treaty (NPT)** and maintaining a balanced investment between civilian and military programs. These nations recognize that, when used properly, nuclear energy can bring immense benefits to their populations by contributing to energy security, industrial modernization, and scientific research.

On the other hand, some nations use nuclear development as a tool of power and intimidation, diverting resources to the construction of arsenals at the expense of investments in infrastructure, education, and healthcare. These countries often operate under secret agendas, concealing their nuclear progress and defying international oversight bodies such as the **International Atomic Energy Agency (IAEA)**.

The relationship between investments in civilian and military nuclear programs is a clear indicator of a country's commitment to progress and global stability. The world's most influential nuclear powers—such as the **United States, Russia, China, France, and the United Kingdom**—allocate significant resources to both the civilian and military sectors, maintaining a balance between security and development. In contrast, countries considered outcasts in the international arena—such as **North Korea and other authoritarian regimes**—prioritize nuclear militarization at the expense of economic growth and the well-being of their populations.

Another crucial point is that countries investing heavily in peaceful nuclear energy generally maintain high standards of transparency and regulation, actively collaborating with international bodies to ensure their activities are safe and well-monitored. In contrast, regimes seeking to develop nuclear weapons clandestinely resort to a lack of transparency, concealment of facilities, and violations of international agreements.

It is, therefore, clear that nuclear energy, in itself, does not pose a threat to humanity. The real danger lies in how this technology is used and the intentions of the governments that control it. The future of global security depends on maintaining a healthy balance between the peaceful use of nuclear energy and the containment of nuclear weapons proliferation. The international community must continue to strengthen oversight mechanisms, promote progressive disarmament, and encourage the responsible development of nuclear energy for the benefit of all humanity.

Table 13: Sources Consulted in Chapter 3

Source	Description
International Atomic Energy Agency (IAEA)	Reports on medical and industrial applications of nuclear energy.
World Nuclear Association	Information on research reactors and non-energy nuclear applications.
National Cancer Institute (Brazil)	Nuclear medicine applications in cancer diagnosis and treatment.
World Health Organization (WHO)	Data on radiotherapy and diagnostic imaging.
NASA	Applications of nuclear energy in probes and space missions.

International Atomic Energy Agency	Publications on nuclear techniques in agriculture and food preservation.
EURATOM	European research initiatives on peaceful nuclear applications.
Scientific Publications	Lancet, Journal of Nuclear Medicine, Physics Today – Studies on civilian use of nuclear energy.
CNEN – National Nuclear Energy Commission (Brazil)	Data on nuclear medicine and radioisotopes.
Scientific and technical outreach books	On peaceful uses of nuclear energy.

Next Chapter: Peaceful Applications of Nuclear Energy – Electricity, Medicine, and Space Exploration

After examining the military sector, we now return to the peaceful applications of nuclear energy. In the next chapter, we will explore in depth how nuclear technology is applied to everyday life—and how it can support economic development and human well-being in a meaningful and sustainable way.

Chapter 4 –
Peaceful Applications of Nuclear Energy: Electricity, Medicine, and Space Exploration

Nuclear energy has played an essential role in the development of modern society, offering innovative solutions to energy, medical, and technological challenges. Despite its association with nuclear weapons, the peaceful application of nuclear technology has provided significant benefits to millions of people around the world.

In this chapter, we will examine the primary areas where nuclear energy is utilized for peaceful purposes, including electricity generation, nuclear medicine, space exploration, and various industrial and scientific applications. These applications demonstrate that nuclear energy can be a powerful tool for human progress when used responsibly and under proper regulations.

Nuclear Energy for Electricity Generation

Electricity generation through nuclear energy is one of its most well-known and widespread applications. Today, dozens of countries operate nuclear power plants to produce reliable, low-carbon electricity.

Global electricity demand has grown exponentially due to population growth, economic development, and the digitalization of modern societies. In this context, nuclear energy has played a crucial role in the energy mix of many nations, ensuring a stable electricity supply with minimal carbon emissions.

Nuclear energy has become one of the world's leading sources of electricity, offering a stable and low-emission alternative to meet the growing global energy demand. Based on the principle of nuclear fission, nuclear power plants generate heat to produce electricity with high efficiency and reliability.

How Nuclear Power Plants Work

Nuclear power plants operate based on the controlled fission of heavy atoms, mainly **uranium-235** and, to a lesser extent, **plutonium-239**. This process occurs inside the nuclear reactor, where atomic nuclei split and release a large amount of energy in the form of heat. This heat is used to generate steam, which drives turbines connected to electrical generators.

The operation of a nuclear power plant can be summarized in the following steps:

1. **Nuclear fission in the reactor core**: Neutrons bombard uranium-235 atoms, causing them to split and release more neutrons, sustaining a controlled chain reaction.

2. **Heat transfer**: The heat generated by fission heats a coolant (typically high-pressure water), preventing reactor overheating.

3. **Steam generation**: The coolant transfers heat to a secondary circuit, where water turns into steam.

4. **Turbine movement**: The pressurized steam spins turbines connected to electric generators.

5. Condensation and recirculation: The steam is cooled and converted back into water, which is recirculated in the system.

The most common types of reactors used for electricity generation include:

- **Pressurized Water Reactor** (PWR): The most widely used globally, operating with pressurized water to cool the core and transfer heat.

- **Boiling Water Reactor** (BWR): Uses water that boils directly in the core to produce steam and drive turbines.

- **High-Temperature Gas Reactor** (HTGR): Uses helium gas as a coolant, operating at higher temperatures with improved efficiency.

- **Molten Salt Reactor** (MSR): An emerging technology that uses liquid salts to enhance safety and efficiency.

Nuclear power plants stand out for their high energy density—meaning a small amount of nuclear fuel generates vast amounts of electricity—enabling continuous operation for months or even years without refueling. This translates to very low operational costs (OPEX), allowing them to produce electricity at competitive prices.

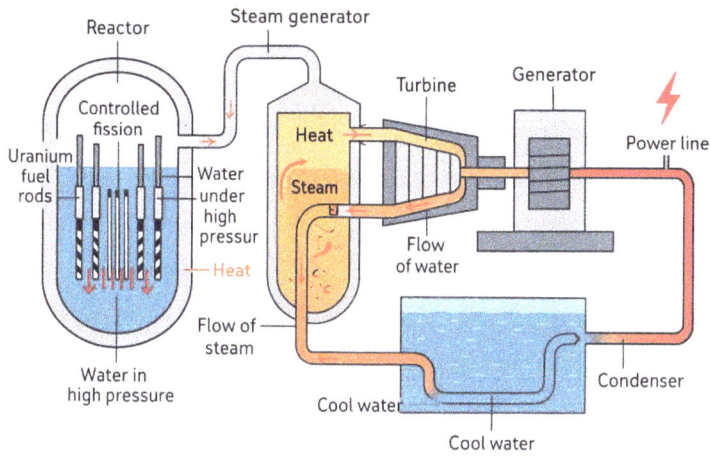

Detailed diagram of a nuclear reactor in operation, showing the main components and the flow of heat and steam in the electricity generation process.

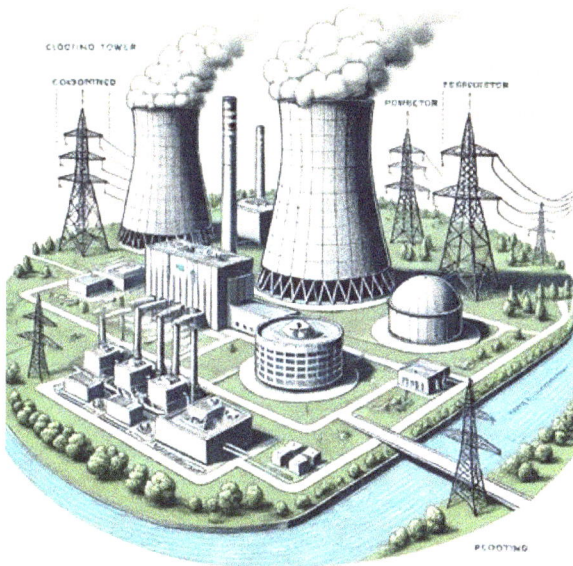

Illustration of a nuclear power plant in operation.

Comparison with Other Energy Sources

Nuclear energy has both advantages and disadvantages compared to other forms of electricity generation.

Table 14: Energy Sources Comparison

Energy Source	CO_2 Emissions	Reliability	Energy Density	Long-term Cost
Nuclear	Very Low	Very High	Very High	Medium
Coal	Very High	High	Medium	Low
Natural Gas	Medium	High	Medium	Medium
Hydropower	Very Low	Medium	High	High
Wind	None	Low	Low	Medium
Solar	None	Low	Low	High

Source: Own elaboration based on the data presented in the Summary Table at the end of this chapter

Benefits, Challenges, and Global Examples of Nuclear Energy in Electricity Generation

Nuclear energy stands out for its reliability (not dependent on weather variations), low greenhouse gas emissions, high energy density, and very low operating costs. However, it requires high upfront investments, strict regulatory oversight, and careful management of nuclear waste.

Renewable sources such as solar and wind offer advantages in sustainability but face issues with intermittency (reliant on sun and wind) and require storage solutions.

Fossil fuels (coal and natural gas) remain widely used due to their low initial costs but have severe environmental impacts, including CO_2 emissions and air pollution.

Benefits:

- **Very low direct environmental impact**: No CO_2 emissions during electricity generation.

- **High reliability**: Plants operate 24/7, ensuring a stable supply.

- **Low operating costs**: Due to the small amount of "nuclear fuel" required, the main expenses are related to highly specialized personnel—engineers, operators, and safety and quality staff.

- **Low land use**: Requires less space than solar and wind farms for equivalent energy production.

- **Lower impact on biodiversity**: Unlike hydroelectric plants, it does not alter aquatic ecosystems.

Challenges:

- **Radioactive waste management**: Waste must be securely stored for long periods.

- **High initial costs**: Building new plants is expensive and time-consuming due to regulatory demands. A substantial portion of these costs is related to the education and training of technical staff—an investment that strengthens national human capital.

- **Risk of accidents**: While extremely rare, incidents like Chernobyl and Fukushima have affected public perception.

- **Political and social issues**: Popular fear and lack of consensus hinder broader acceptance in some countries.

Despite these challenges, new technologies such as Small Modular Reactors (SMRs) and passive safety systems are making nuclear energy more viable and safer for the future.

Examples of Countries Heavily Dependent on Nuclear Energy

Several countries rely on nuclear energy as their main source of electricity. The most notable example is France, where approximately 70% of electricity comes from nuclear reactors.

France FR

- Operates 56 nuclear reactors.

- Produces electricity at a relatively low cost.

- Exports energy to neighboring countries.

Other countries with high nuclear dependency include:

- **Slovakia SK** → 53% of electricity from nuclear reactors.

- **Ukraine UA** → 51% nuclear-generated electricity.

- **Hungary HU** → 49% of electricity from nuclear.

- **Belgium BE** → 47% of electricity from nuclear energy.

In the United States, Japan, and Russia, nuclear accounts for about 20% to 30% of the energy mix, while countries like Germany and Italy have been reducing their use for political reasons.

Table 15: Table of Countries with Nuclear Reactors

Country	Number of Reactors	Installed Capacity (MW)
United States	93	95523
France	56	61370
China	54	52200
Russia	37	27727
Japan	33	31679
South Korea	25	24429
Canada	19	13624
Ukraine	15	13107
United Kingdom	9	5923
Sweden	6	6927
India	22	6885
Germany	6	8113
Belgium	7	5942
Spain	7	7121
Czech Republic	6	3932
Finland	5	4400
Switzerland	4	2960

Hungria	4	1902
Slovakia	4	1814
Bulgaria	2	1926
Brazil	2	1884
South Africa	2	1860
Mexico	2	1552
Romania	2	1300
Argentina	3	1641
Iran	1	1020
Armenia	1	375
Netherlands	1	482
Pakistan	6	2332
United Arab Emirates	4	5600

Source: Own elaboration based on the data presented in the Summary Table at the end of this chapter

Note: The data above has been compiled from various sources, including the International Atomic Energy Agency and the World Nuclear Association. Figures may vary with updates and the commissioning of new reactors.

Nuclear energy is an essential technology for a stable, sustainable, and cost-effective energy supply—particularly in a world seeking to reduce carbon emissions. Despite technical and societal challenges, technological advancements and the increasing need for clean energy sources continue to drive its growing global importance.

Its low carbon emissions play a decisive role in climate change mitigation. The 24/7 operational capacity of nuclear plants ensures stability and an unmatched ability to manage power grids efficiently.

Another key aspect is energy independence: countries with nuclear reactors significantly reduce their exposure to fossil fuel supply shocks. Finally, nuclear power plants offer long-term efficiency and longevity—operating between 40 and 60 years— which ensures the viability of long-term investments.

Table 16: Types of Nuclear Reactors for Power Generation

Reactor Type	Fuel Used	Key Characteristics
PWR (Pressurized Water Reactor)	Enriched uranium	Most common worldwide; high-pressure water.
BWR (Boiling Water Reactor)	Enriched uranium	Boils water directly in the reactor core.
MSR (Molten Salt Reactor)	Thorium or fuel dissolved in salt	High efficiency and safety; emerging technology.

Source: Own elaboration based on the data presented in the Summary Table at the end of this chapter

Nuclear Medicine and Radiotherapy

Nuclear medicine is one of the most revolutionary applications of nuclear energy, enabling accurate diagnoses and effective treatments for a wide range of diseases, including cancer. It is based on the use of radioisotope elements that emit radiation and can be used for therapeutic or diagnostic purposes.

The ability to visualize internal organs in real time and treat tumors with high precision has made nuclear medicine essential to modern medical practice. Advances in this field have led not only to a better understanding of diseases but also to more effective and less invasive therapies.

Use of Radioisotopes for Diagnosis and Treatment:

Radioisotopes are unstable atoms that emit radiation as they decay into more stable forms. This radiation can be used to detect abnormalities in the body or precisely destroy diseased cells.

Table 17: Medical Radioisotopes

Radioisotope	Application	Half-life
Technetium-99m (^{99m}Tc)	Imaging diagnostics (heart, bones, kidneys)	6 hours
Iodine-131 (^{131}I)	Treatment of thyroid cancer and hyperthyroidism	8 days
Fluorine-18 (^{18}F)	PET scan for oncology and neurology	110 minutes
Cobalt-60 (^{60}Co)	Radiotherapy against cancer	5.3 years
Gallium-67 (^{67}Ga)	Infection diagnostics	79 hours
Thallium-201 (^{201}Tl)	Cardiac studies	73 hours

Source: Own elaboration based on the data presented in the Summary Table at the end of this chapter

Each of these radioisotopes has specific characteristics that make them suitable for different types of scans or treatments.

The column "**Half-Life**" represents the time it takes for **half of the atoms of a radioisotope to decay** into a more stable element, emitting radiation in the process.

Table 18: Medical Radioisotopes and Their Applications

Radioisotope	Medical Use	Type of Radiation
Technetium-99m	Diagnostic imaging (gamma camera)	Gamma
Iodine-131	Thyroid cancer treatment	Beta and gamma
Fluorine-18	PET scan – functional imaging	Positron
Cobalt-60	Radiotherapy for tumors	Gamma

Source: Own elaboration based on the data presented in the Summary Table at the end of this chapter

Why is half-life important in nuclear medicine?

- **It determines the duration of the radioisotope's effect**:

  - Radioisotopes with a short half-life (such as **Fluorine-18**, used in PET scans) quickly disappear from the body, reducing radiation exposure.

  - Radioisotopes with a long half-life (such as **Cobalt-60**, used in radiotherapy) can be stored and used for years.

- **It helps adjust dosage for diagnostics and treatments**:

  - If a radioisotope decays too quickly, higher doses may be required.

  - If the half-life is too long, the material may stay in the body longer than needed.

- **It impacts waste storage and disposal**:

  - Radioisotopes with very long half-lives require safe storage for decades or even centuries, depending on the application.

Examples:

- **Technetium-99m (^{99m}Tc)**, with a half-life of only **6 hours**, is ideal for medical imaging because it quickly clears from the body.

- **Iodine-131 (^{131}I)**, with a half-life of **8 days**, is used in thyroid treatment because it remains active long enough to destroy diseased cells.

- **Cobalt-60 (^{60}Co)**, with a half-life of **5.3 years**, is excellent for radiotherapy because it can be stored for long periods without losing effectiveness.

Chart 27: Radioactive Lifecycle and Half-Life

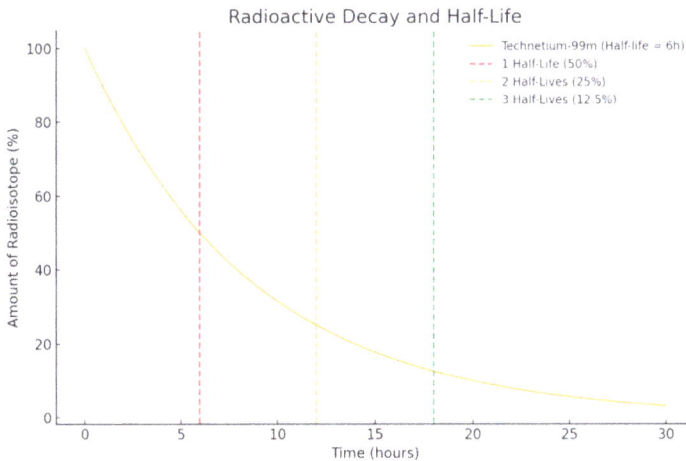

Source: Own elaboration based on the data presented in the Summary Table at the end of this chapter

An explanatory chart on half-life, illustrating how the amount of a radioisotope (e.g., Technetium-99m) decreases over time:

- After **one half-life (6 hours)** → **50%** of the material remains.

- After **two half-lives (12 hours)** → **25%** remains.

- After **three half-lives (18 hours)** → **12.5%** remains.

- And so on, following an **exponential decay** curve.

This concept is crucial in nuclear medicine, as it determines the optimal timing for scans and treatments.

PET Scans, Cancer Radiotherapy, and Sterilization of Medical Equipment

Nuclear medicine has three major application areas:

1. Diagnostic Imaging: PET Scans and Scintigraphy

Nuclear imaging allows real-time visualization of organ function, something impossible with conventional X-rays.

- **PET Scan (Positron Emission Tomography)**

  - Uses **Fluorine-18** linked to a glucose molecule (**FDG**). Since cancer cells consume more glucose, the tracer accumulates in those areas, enabling early tumor detection.

  - Also used to assess neurological diseases such as Alzheimer's and epilepsy.

- **Scintigraphy**

  - Uses **Technetium-99m** and other isotopes to examine organs such as the heart, bones, kidneys, and lungs.

  - Helps assess blood flow and kidney function and detect hidden bone fractures.

These methods are less invasive than biopsies and allow for early diagnoses, increasing treatment success rates.

2. Cancer Radiotherapy

Radiotherapy is one of the most effective ways to treat cancer, using radiation to destroy tumor cells.

Main radiotherapy modalities:

- **External Radiotherapy**
    - Equipment like linear accelerators directs radiation beams at the tumor.
    - **Cobalt-60** and particle accelerators are widely used.

- **Brachytherapy**
    - Radioisotopes are inserted inside or near the tumor, releasing radiation directly into cancer cells.
    - Used for prostate, uterine, and breast cancer.

- **Radionuclide Therapy**
    - **Iodine-131** for thyroid cancer.
    - **Lutetium-177** for neuroendocrine tumors.

The great advantage of radiotherapy is its high precision, minimizing damage to healthy tissue around the tumor.

3. Sterilization of Medical Equipment and Blood Transfusions

Radiation is also used to sterilize medical materials and ensure the safety of hospital supplies.

- **Cobalt-60** is used to sterilize syringes, gloves, catheters, and prostheses, eliminating viruses and bacteria.

- Blood irradiation prevents graft-versus-host disease, common in immunosuppressed patients after transfusions.

This technique allows sterilization without heat or chemicals, preserving the integrity of the materials.

Safety and Regulation in the Medical Field

The use of radioisotopes in medicine requires strict safety protocols to protect both patients and healthcare professionals.

1. International Regulation

Safety in nuclear medicine is governed by organizations such as:

- **International Atomic Energy Agency (IAEA)** – Establishes global safety standards.

- **International Commission on Radiological Protection (ICRP)** – Sets dose limits for professionals and patients.

- **National authorities** such as the FDA (USA), CNEN (Brazil), and ASN (France) regulate the clinical use of radioisotopes.

2. Protection of Patients and Professionals

- **Dose monitoring**: Patients receive the lowest possible dose to minimize risk.

- **Worker protection:** Equipment such as dosimeters and shielding reduce exposure.

- **Safe storage**: Radioisotopes are handled in shielded rooms with rigorous protocols.

Radiation used in nuclear medicine is safe when applied correctly, and its benefits outweigh the risks when compared to conventional exams and treatments.

Chart 28 : Relationship between Life Expectancy and Nuclear Medicine

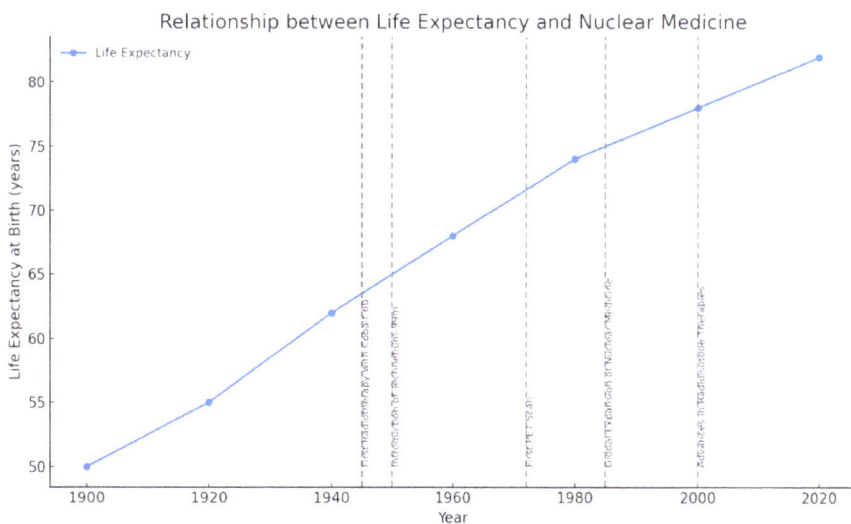

Source: Own elaboration based on the data presented in the Summary Table at the end of this chapter

Chart shows the relationship between increased life expectancy in developed countries and the advancement of nuclear medicine.

What does it show?

- The blue line represents the increase in life expectancy throughout the 20th and 21st centuries.

- The gray vertical lines mark key events in nuclear medicine, such as the discovery of X-rays, the introduction of Technetium-99m, and advances in PET scans and radiotherapy.

- The graph highlights a consistent rise in life expectancy following the implementation of nuclear-based medical technologies.

The Importance of Illustrating the Link Between Life Expectancy and Nuclear Medicine:

While it is challenging to establish a direct correlation due to the influence of multiple factors on longevity, we can present data that contextualizes this relationship.

Life Expectancy Trends in Developed Countries:

Life expectancy in developed countries increased significantly throughout the 20th century. For example:

- **1950s**: Life expectancy at birth was approximately 68 years.

- **1980s**: This rose to about 74 years.

- **2020s**: Life expectancy reached approximately 80 years.

140

Development of Nuclear Medicine:

Nuclear medicine reached important milestones that contributed to advances in disease diagnosis and treatment:

1. **1895**: Discovery of X-rays by Wilhelm Röntgen, marking the beginning of radiology.

2. **1950s**: Development of the Technetium-99m generator, enabling production of radioisotopes for medical diagnostics.

3. **1970s**: Advances in magnetic resonance imaging (MRI), with contributions from researchers like Peter Mansfield.

To illustrate the relationship between life expectancy and advances in nuclear medicine, a line chart was created with the following elements (Chart 28):

- X-Axis: Timeline (years), from 1900 to 2020.

- Y-Axis: Life expectancy at birth (in years).

- Line: Life expectancy trend in developed countries.

- Historical markers: Key points on the timeline indicating significant discoveries and implementations in nuclear medicine.

It is important to note that increases in life expectancy result from a combination of factors, including improvements in nutrition, sanitation, vaccinations, medical treatments, and technological advancements. Nuclear medicine has played a crucial role, especially in early diagnosis and effective treatment of serious illnesses, contributing to reduced mortality and improved quality of life.

Methodological Note and Limitations of the Analysis:

It is important to clarify that the relationship presented in this chapter between increased life expectancy and nuclear medicine does not necessarily imply a direct cause-and-effect correlation. Longevity growth in developed countries over the 20th and 21st centuries has been driven by a series of interrelated factors, including improved nutrition, mass vaccination, the development of antibiotics, surgical advances, greater access to public health, and progress in basic sanitation.

Nevertheless, nuclear medicine has played an undeniable role in revolutionizing the diagnosis and treatment of serious diseases such as cancer and cardiovascular and neurological disorders. It has enabled early detection, more effective therapies, and higher precision in medical procedures. These technologies have contributed to a better quality of life and reduced mortality, making nuclear medicine an essential tool in modern healthcare.

The chart should be interpreted as a **speculative analysis** based on historical events and general trends and is not intended to establish a rigid statistical correlation. Its purpose is to demonstrate how medical innovations, including nuclear medicine, are part of a broader set of advances that have contributed to the extension of human longevity.

Nuclear medicine has revolutionized the diagnosis and treatment of diseases. From advanced imaging to effective cancer therapies, its impact on healthcare is profound. The continued advancement of nuclear medical technologies promises faster diagnoses, more effective treatments, and greater safety for both patients and professionals.

Space Exploration and Nuclear Energy

Space exploration has always been linked to the search for reliable and efficient energy sources. In the vacuum of space, where there is no oxygen for combustion and sunlight is limited, nuclear energy has emerged as a viable solution to power spacecraft, probes, and even future lunar and Martian bases.

From the first experiments in the 1960s to today's most ambitious projects—such as those by NASA and SpaceX—nuclear reactors and Radioisotope Thermoelectric Generators (RTGs) have become essential for long-duration space missions.

Currently, with renewed interest in interplanetary exploration—especially the colonization of Mars, one of the main goals of Elon Musk and SpaceX—nuclear energy is once again a central topic for enabling faster space travel and the creation of sustainable infrastructure beyond Earth.

Nuclear Reactors in Space

Unlike solar energy, which loses efficiency the farther we get from the Sun, nuclear energy can provide consistent and reliable power, making it essential for long-duration missions. Space nuclear reactors have been developed to generate electricity and heat in extreme environments.

The First Nuclear Reactor in Space: SNAP-10A

The SNAP-10A was the first nuclear reactor launched into space, deployed by the United States in 1965.

- Developed by the Atomic Energy Commission (AEC) and the Air Force Systems Command.

- Generated 500 watts of electricity using enriched uranium as fuel.

- Operated for 43 days before shutting down due to an electrical failure.

The SNAP-10A demonstrated that nuclear energy was viable in space, although it was never followed by a new generation of operational reactors.

SNAP-10A (Space Nuclear Reactor)

The SNAP-10A was the first U.S. nuclear reactor launched into space in 1965. Images and technical details can be found in the article by the World Nuclear Association.

The Revival of Space Reactors: The Kilopower Project

In recent years, NASA and other agencies have resumed investments in space nuclear reactors. One of the most promising projects is Kilopower, developed by NASA in partnership with the U.S. Department of Energy.

Key Features of Kilopower:

- Generates between 1 and 10 kW of electricity.

- Powered by Uranium-235 and uses Stirling converters to produce energy.

- Can operate continuously for over 10 years without maintenance.

- Designed for use on lunar and Martian bases, providing electricity for habitats and scientific equipment.

NASA successfully tested a Kilopower prototype, named KRUSTY, in 2018. The goal is to use this type of technology to enable astronauts to live and work on Mars, where solar energy may be insufficient during dust storms.

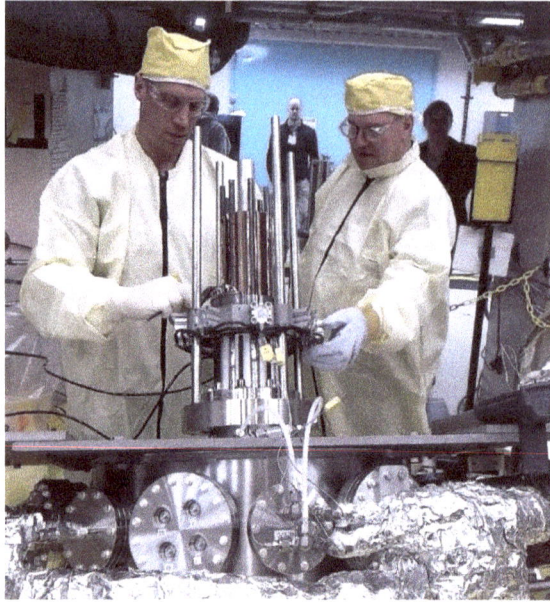

Kilopower (NASA Nuclear Reactor Project)

Kilopower is a recent NASA project designed to provide power for long-duration space missions.

Table 19: Nuclear Power Sources for Space Exploration

Technology	Space Applications	Example Mission
RTG (Radioisotope Thermoelectric Generator)	Continuous power for probes and rovers	Voyager, Cassini, Curiosity
SNAP-10A	Experimental space reactor (1965)	US satellite test
Kilopower (KRUSTY)	Compact reactor for lunar/martian bases	Prototype successfully tested in 2018

Source: Own elaboration based on the data presented in the Summary Table at the end of this chapter

SpaceX, Mars, and the Use of RTGs in Space Missions

Connection with SpaceX and Mars:

Elon Musk frequently emphasizes the importance of a reliable energy source for the colonization of Mars. Although SpaceX focuses primarily on rocket development, the company has collaborated with NASA on studies regarding infrastructure for Martian bases, where Kilopower could play a key role in providing power.

Use of RTGs (Radioisotope Thermoelectric Generators) in Probes and Rovers:

In addition to nuclear reactors, another major advancement in space energy has been the development of Radioisotope Thermoelectric Generators (RTGs).

How do RTGs work?

RTGs use the radioactive decay of Plutonium-238 to generate heat, which is then converted into electricity through a thermoelectric process.

Advantages:

- Extremely reliable, capable of operating for decades without maintenance.

- Resistant to extreme conditions, such as intense cold and cosmic radiation.

- Essential for missions in regions where solar power is not viable (e.g., Jupiter and beyond).

Famous missions that used RTGs:

Table 20: Nuclear-Powered Space Missions

Mission	Launch Year	Destination	Operational Time
Voyager 1 and 2	1977	Interstellar Space	Still operational
Cassini-Huygens	1997	Saturn	20 years
Curiosity Rover	2011	Mars	Still operational
Perseverance Rover	2020	Mars	Still operational

Source: Own elaboration based on the data presented in the Summary Table at the end of this chapter

RTGs enabled probes like Voyager 1 and 2, launched in the 1970s, to continue transmitting data from interstellar space more than 45 years later!

Curiosity Rover (Mars Exploration)

The Curiosity rover has been exploring Mars since 2012. Images and mission updates can be found on NASA's official website.

Connection with SpaceX and Future Mars Missions:

The Perseverance and Curiosity rovers, currently exploring Mars, rely on RTGs for power. If Elon Musk and SpaceX succeed in establishing a human colony on Mars, RTGs and nuclear reactors will be essential to provide energy for the first human settlements.

Future Applications: Nuclear Propulsion for Interplanetary Travel

Future space exploration demands faster and more efficient travel solutions. Nuclear propulsion emerges as one of the most promising alternatives.

Types of Nuclear Propulsion for Space:

- **Nuclear Thermal Propulsion** (NTP)

    o Uses a nuclear reactor to heat liquid hydrogen, which is then expelled to generate thrust.

    o Can cut travel time to Mars in half compared to traditional chemical rockets.

    o NASA tested concepts such as NERVA in the 1960s and resumed research through the DRACO project in partnership with DARPA.

- **Nuclear Electric Propulsion** (NEP)

    o Uses a reactor to generate electricity that powers ion engines.

    o More fuel-efficient, ideal for long-duration missions.

    o Successfully tested in missions such as the Deep Space 1 probe.

SpaceX and Plans for Mars:

While SpaceX currently uses chemical rockets such as the Starship, nuclear propulsion may be a key technology to reduce travel time and make interplanetary missions safer. The company has expressed interest in collaborating with NASA on future research into this technology.

Nuclear energy in space has been vital for exploring distant planets, powering probes and rovers, and could now become a cornerstone of human colonization of Mars.

With the growing ambition of companies like SpaceX and Blue Origin—alongside NASA and other space agencies—nuclear energy is once again at the center of discussions on interplanetary travel and space colonization.

The question is no longer "if" but "when" we will see nuclear reactors operating on Mars and nuclear-powered rockets carrying humans beyond our solar system.

Other Industrial and Scientific Applications

While nuclear energy is widely recognized for its role in electricity generation and medicine, its impact extends to many other areas. Agriculture, industry, archaeology, and geology are fields where the application of nuclear techniques has led to significant advances, improving productivity, safety, and our understanding of Earth's past.

Thanks to the use of radioisotopes and nuclear techniques, processes that were once impossible or highly imprecise have become efficient and reliable, driving technological development and sustainability across multiple sectors.

Nuclear Applications in Agriculture

Nuclear energy plays a crucial role in food security, increasing agricultural productivity and reducing post-harvest losses. The main applications include food irradiation and induced mutation for genetic improvement.

Food Irradiation:

Food irradiation is a technique that uses ionizing radiation to eliminate bacteria, fungi, and parasites, extending shelf life without compromising nutritional value.

How does it work?

Food is exposed to beams of gamma rays (**Cobalt-60** or **Cesium-137**), X-rays, or electron beams. This exposure destroys harmful microorganisms without making the food radioactive.

Advantages:

- Eliminates disease-causing microorganisms (e.g., Salmonella, E. coli).

- Increases food shelf life without the need for chemical preservatives.

- Reduces pesticide use, minimizing environmental impact.

Common irradiated products:

- Fruits and vegetables (to prevent pests and delay ripening).

- Meat and seafood (to eliminate bacteria).

- Grains and spices (to eliminate insects and fungi).

The World Health Organization (WHO) and the International Atomic Energy Agency (IAEA) recognize food irradiation as a safe and beneficial process for public health.

Induced Mutation for Genetic Improvement:

Radioisotopes are also used to induce beneficial genetic mutations in plants, accelerating the development of more productive and pest-resistant varieties.

How does it work?

Seeds or plant tissues are exposed to gamma rays or neutrons, inducing DNA mutations. Beneficial mutations are selected and propagated to develop new crop varieties.

Benefits:

- Development of plants resistant to diseases and pests.

- Reduction in the use of agrochemicals.

- Increased agricultural output to fight global hunger.

Success Example:

The International Rice Research Institute (IRRI) used this technique to develop rice varieties resistant to floods and droughts, contributing to food security in Asia.

The International Rice Research Institute (IRRI) is an independent, non-profit organization focused on agricultural research and training in rice cultivation. Founded in 1960 by the Ford and Rockefeller Foundations in collaboration with the Philippine government, IRRI's mission is to reduce poverty and hunger, improve the health of rice farmers and consumers, and promote environmental sustainability in rice production.

Headquartered in Los Baños, Philippines, IRRI operates in 17 countries and is renowned for its role in the Green Revolution of the 1960s and 1970s—particularly through the development of high-yield rice varieties like IR8, which helped prevent food crises in several Asian regions.

IRRI has utilized nuclear techniques, such as mutation induction through radiation, to develop new rice varieties that are more productive and resistant to adverse conditions. These techniques have enabled the creation of crops that contribute to increased agricultural productivity and food security.

Chart 29: Impact of Nuclear Technology on Rice Production

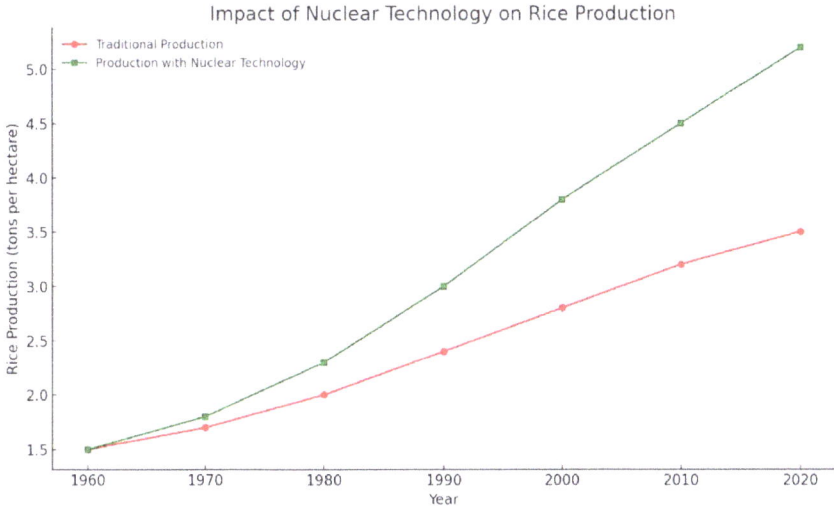

Impact of Nuclear Technology on Rice Production

Source: Own elaboration based on the data presented in the Summary Table at the end of this chapter

Illustrative chart comparing rice production before and after the application of nuclear technology

What does it show?

- The red line represents traditional rice production over the decades.

- The green line shows the impact of radiation-induced genetic improvement (nuclear technology), which led to a significant increase in productivity.

Starting in the 1970s–1980s, when improved rice varieties began to be widely cultivated (thanks to projects like IRRI), there was a major leap in yield per hectare.

Today, more than 3,000 plant varieties have been developed using this technique, contributing to sustainable agriculture in various countries.

Applications in Industry

Modern industry relies heavily on nuclear energy to ensure quality, safety, and efficiency in various processes. Techniques such as industrial radiography, thickness measurement, and quality control use radioisotopes to detect flaws that are invisible to the naked eye.

Detection of Material Defects:

Industrial radiography is a non-destructive method used to inspect the integrity of metal structures, welds, and mechanical components.

How does it work?

- Gamma rays (Cobalt-60 or Iridium-192) are directed at the object to be inspected.

- A detector or radiographic film captures the internal image, revealing cracks, air pockets, and structural imperfections.

Applications:

- **Aérospatiale** → Inspection of turbines and aircraft fuselages.

- **Oil and Gas** → Verification of the integrity of pipelines and welds.

- **Civil Construction** → Inspection of bridge and building structures.

The advantage of nuclear radiography is its ability to detect structural flaws early, helping to prevent accidents and ensure the safety of critical industrial operations.

Ometto Industrial Radiography System

Ometto provides fixed industrial radiography equipment used for the inspection of welds, cast parts, and metal structures, ensuring the quality and integrity of materials.

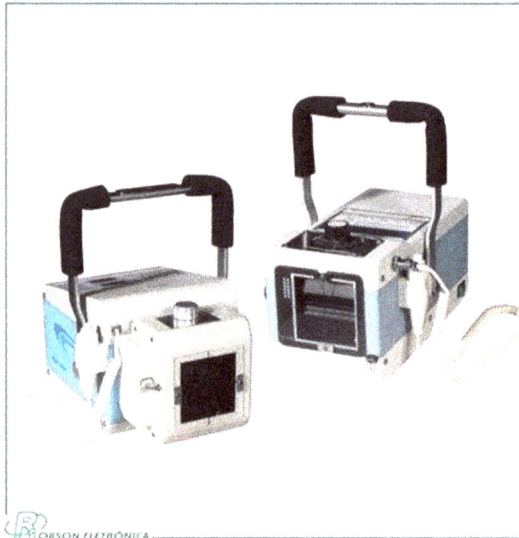

Julio Verne Industrial X-Ray Portable Radiography System

Portable equipment that emits ionizing radiation (X-rays or gamma rays) through the part being inspected, allowing the detection of defects or cracks in the material of the parts.

Important Considerations:

- **Safety**: The use of industrial radiography equipment requires specialized training, certification, and strict safety measures to protect operators and the environment from potential radiation exposure.

- **Applications**: This equipment are widely used for weld inspection, detection of material defects, quality control in manufacturing processes, and preventive maintenance in various industrial sectors.

Thickness Measurement and Quality Control:

Thickness measurement using radioisotopes is widely used to ensure quality in manufacturing processes.

How does it work?

- Beta radiation beams (Strontium-90) or gamma rays are emitted through the material.

- Sensors detect the amount of radiation absorbed, determining the exact thickness of the product.

Industrial Uses:

- **Paper Manufacturing** → Thickness control of sheets.

- **Steel Production** → Measurement of metal sheets.

- **Automotive Industry** → Quality control in tires and metal parts.

This technology helps reduce waste, improve manufacturing accuracy, and ensure compliance with international standards.

Applications in Archaeology and Geology

Nuclear techniques play a crucial role in understanding Earth's history and human civilizations, enabling accurate analysis of ancient materials and geological processes.

Carbon-14 Dating:

Carbon-14 dating is a fundamental technique in archaeology for determining the age of organic materials such as wood, bones, and tissues, up to approximately 50,000 years.

Principle of the Method:

- **Incorporation of Carbon-14**: During life, organisms absorb carbon, including the radioactive isotope Carbon-14 (^{14}C), which is present in the atmosphere.

- **Decay after Death**: After death, the absorption of ^{14}C ceases, and the isotope begins to decay with a half-life of approximately 5,730 years.

- **Age Calculation**: By measuring the remaining amount of ^{14}C in the material, it is possible to estimate the time elapsed since the organism's death.

Dating Process:

1. **Sample Collection**: Carefully extracting a portion of the material to be dated.

2. **Sample Preparation**: Cleaning and chemical treatment to remove contaminants.

3. **Radioactivity Measurement**: Using mass spectrometry or radiation counters to determine the amount of ^{14}C present.

4. **Age Calculation**: Applying mathematical formulas that relate the remaining ^{14}C amount to the time elapsed since the organism's death.

CARBON-14 DATING

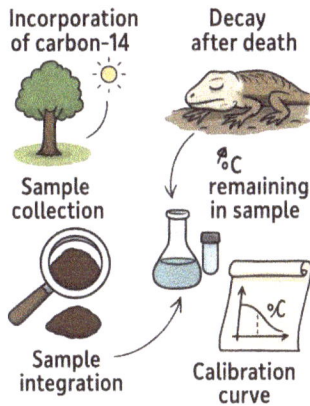

Incorporation of carbon-14

Decay after death

Sample collection

%C remaining in sample

Sample integration

Calibration curve

Studies of the Earth's Crust

In geology, nuclear techniques are employed to analyze the composition and age of rocks, contributing to the understanding of the planet's formation and evolution.

Main Techniques:

- **Uranium-Lead Dating**: Utilizes the decay of Uranium-238 to Lead-206 to determine the age of minerals like zircon, allowing estimates of up to billions of years.

- **Potassium-Argon Dating**: Based on the decay of Potassium-40 to Argon-40, it helps date volcanic rocks.

Uranium-Lead Dating Process:

- **Sample Collection**: Extraction of specific minerals, such as zircon, from rocks.

- **Preparation and Analysis**: Measurement of uranium and lead isotopic ratios using mass spectrometry.

- **Data Interpretation**: Age calculation based on isotopic ratios and known decay rates.

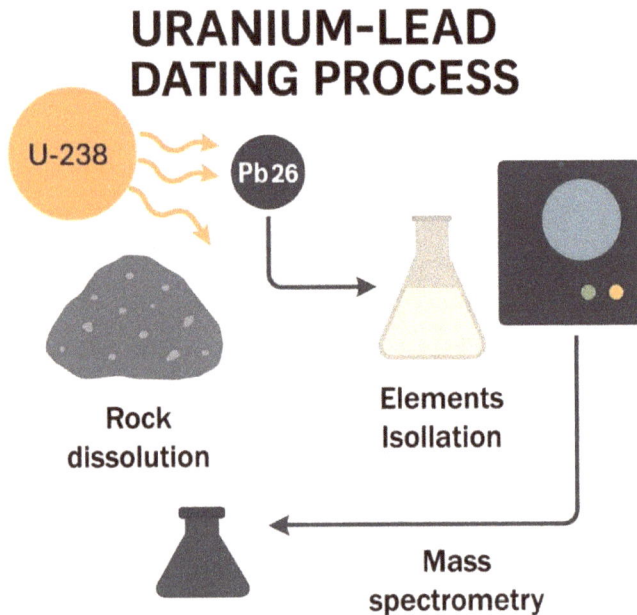

URANIUM-LEAD DATING PROCESS

U-238

Pb 26

Rock dissolution

Elements Isollation

Mass spectrometry

Muon Radiography

Muon radiography

Muon radiography is an innovative technique that uses subatomic particles called muons to investigate the internal structure of large geological and archaeological objects.

Principle of the Method:

- **Muon Penetration**: Muons, produced by the interaction of cosmic rays with the atmosphere, have a high penetration capability through dense materials.

- **Detection of Density Variations**: As they pass through objects, muons lose energy depending on the material's density, allowing the creation of internal images.

Applications:

- **Archaeology**: Detection of hidden chambers in pyramids and other ancient structures.

- **Geology**: Monitoring volcanic activity and detecting underground cavities

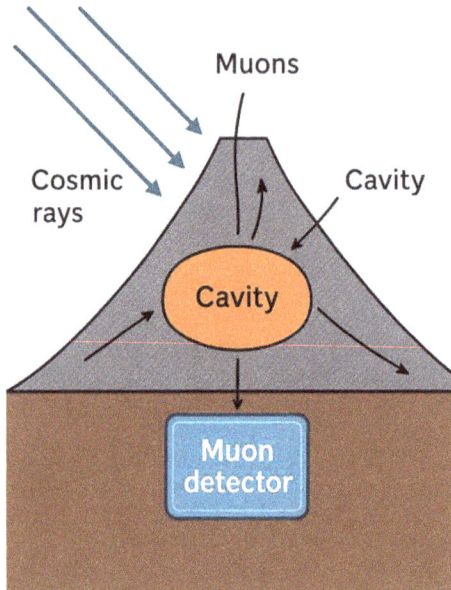

MUON RADIOGRAPHY OF A VOLCANO

Muons

Cosmic rays

Cavity

Cavity

Muon detector

Isotopic Tracers:

This technique involves the substitution of atoms in molecules with their isotopes, allowing the tracking of geological and biochemical processes.

Applications:

- **Study of Biogeochemical Cycles**: Tracking elements like carbon and nitrogen across different environmental compartments.

- **Groundwater Dating**: Using hydrogen and oxygen isotopes to determine the age and origin of groundwater.

SCHEMATIC OF ISOTOPIC TRACING FOR ENVIRONMENTAL STUDIES

Water Vulnerability
Identifying 2-yr or TWV via sources of H_2O

Vegetation Movement
Tracing carbon or nitrogen tracing uptake

Storage and Incineration of Deposits
Studying greenhouse gas release from landfills or facilities

Recycling and Mining
Assessing metal life cycle impacts

Atmosphere Analysis
Carbon sequestration or emissons

Volcanism
Determining past temperature conditions from eruptions

Chapter 4: Conclusion

Nuclear energy, often associated with military uses or electricity generation, plays a crucial role in various sectors of society. Its peaceful applications have transformed the way we diagnose and treat diseases, explore space, ensure food and industrial safety, and study Earth's and humanity's history.

Nuclear medicine has revolutionized the diagnosis and treatment of severe diseases, especially cancer and cardiovascular diseases. The use of radioisotopes in PET scans, radiotherapy, and medical equipment sterilization has led to significant advancements in longevity and quality of life for populations.

Space exploration has become possible largely thanks to nuclear energy. RTGs and space reactors like Kilopower enable long-duration missions, while nuclear propulsion may be the key to future interplanetary travel and Mars colonization.

In agriculture, food irradiation helps reduce waste and ensures food security, while induced mutation allows the development of more resistant and productive crops. In industry, nuclear techniques are essential for material inspection, quality control, and defect detection, ensuring safety and efficiency across various sectors.

Methods like Carbon-14 and Uranium-Lead dating allow for the study of Earth's and humanity's history with accuracy, while muon radiography and isotopic tracers assist in exploring the underground and understanding environmental processes.

With the advancement of technology, nuclear energy will continue to play a crucial role in human progress. Innovations in modular reactors, space propulsion, medical therapies, and environmental monitoring promise new frontiers for this technology, making it increasingly safe and efficient.

When used responsibly, nuclear energy has the potential to improve people's lives, expand our horizons, and drive sustainable development. The challenge of the future will be to maximize its benefits while minimizing risks and ensuring the ethical and safe use of this powerful scientific tool.

Table 21: Other Non-Energy Nuclear Applications

Application Area	Nuclear Use	Benefits
Agriculture	Food irradiation and pest control	Preservation, reduced losses, food safety
Geology	Dating and isotopic tracers	Understanding geological processes and natural cycles
Archaeology	Muon radiography and dating	Exploration without damaging structures
Industry	Quality control, thickness, density	Greater accuracy and process efficiency

Source: Own elaboration based on the data presented in the Summary Table at the end of this chapter

Table 22: Sources Consulted in Chapter 4

Source
International Atomic Energy Agency (IAEA) – Reports on medical and industrial applications of nuclear energy.
World Nuclear Association – Information on research reactors and non-energy uses of nuclear.
National Cancer Institute (INCA) – Applications of nuclear medicine in cancer diagnosis and treatment.
World Health Organization (WHO) – Data on radiotherapy and imaging diagnosis.
NASA – Applications of nuclear energy in probes and space missions.
International Atomic Energy Agency – Publications on nuclear techniques in agriculture and food preservation.
EURATOM – European initiatives for research in peaceful nuclear applications.
Scientific publications (Lancet, Journal of Nuclear Medicine, Physics Today) – Studies on the civil use of nuclear energy.
CNEN – National Nuclear Energy Commission (Brazil) – Data on nuclear medicine and radioisotopes.
Books on scientific dissemination and technical literature on peaceful uses of nuclear energy.

Preparation for Next Chapter: Nuclear Waste Safety and Management – Myths and Solutions

In the next chapter, we will dive into some of the most debated and often misunderstood issues regarding nuclear energy:

- The safety of nuclear power plants is a subject filled with perceptions, often based on historical fears and incomplete information. What does the technical and scientific reality say?

- Nuclear waste is frequently presented as an unsolvable problem. But is this true?

We will also confront the most common myths, such as:

- "Nuclear waste remains dangerous for millions of years."

- "There is no safe solution for storing nuclear waste."

- "It is impossible to prevent new accidents like Chernobyl or Fukushima."

On the other hand, we will explore real solutions and emerging technologies that are revolutionizing the way the industry addresses these issues, including:

- Passive safety systems.

- Next-generation reactors (such as SMRs and fast reactors).

- Definitive solutions for high-level waste, like the Onkalo repository (Finland).

- Effective public communication strategies.

"More than a technical issue, nuclear safety is also a matter of trust, transparency, and responsibility towards future generations."

Chapter 5 –
Nuclear Safety and Waste Management:
Myths and Solutions

Safety has always been one of the central pillars of nuclear energy development. From the early experimental reactors of the 1940s to today's advanced Generation III+ and IV nuclear power plants, nuclear technology has evolved to minimize risks and ensure the protection of both populations and the environment. Nevertheless, despite significant technological advances, nuclear energy continues to carry a strong perception of risk, driven by three main factors:

1. Its association with nuclear weapons and historical accidents – such as Chernobyl (1986) and Fukushima (2011).

2. The fear of radioactivity – often exacerbated by misinformation and scientific misinterpretation.

3. The issue of nuclear waste – is viewed by some as a persistent and unsolvable threat.

Public Fear and Perception of Nuclear Energy

Recent surveys indicate that nuclear energy is often perceived as one of the most dangerous forms of electricity generation despite empirical data suggesting otherwise. According to the International Energy Agency (IEA) and the World Health Organization (WHO), nuclear energy has one of the lowest fatality rates per terawatt-hour (TWh) generated—lower than coal, oil, and even biomass.

A study by Our World in Data shows that between 1965 and 2020, the fatality rate per TWh generated was:

- **Coal**: 24.6 deaths/TWh

- **Oil**: 18.4 deaths/TWh

- **Natural Gas**: 2.8 deaths/TWh

- **Hydropower**: 1.3 deaths/TWh

- **Solar**: 0.02 deaths/TWh

- **Nuclear**: 0.007 deaths/TWh

Chart 30: Fatality Rate by Energy Source

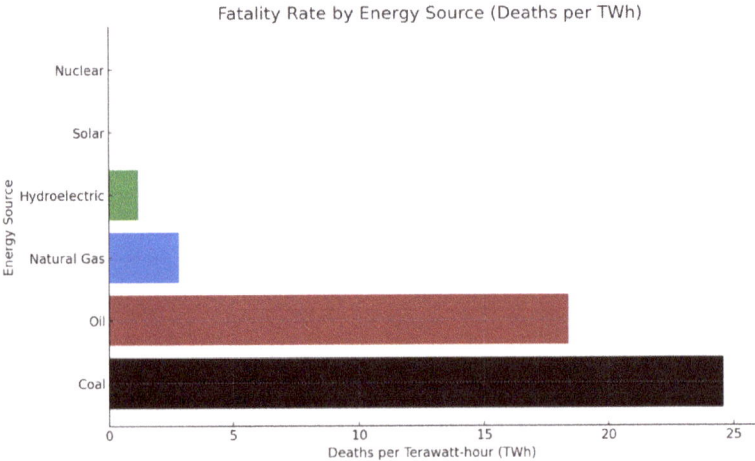

Fatality Rate by Energy Source (Deaths per TWh)

Source: Own elaboration based on the data presented in the Summary Table at the end of this chapter

These figures demonstrate that nuclear energy is one of the safest forms of electricity generation. However, major events such as the Chernobyl and Fukushima accidents have reinforced in the collective imagination the idea that nuclear energy is inherently dangerous.

Another factor that intensifies public concern is radioactivity. Many people confuse the concept of controlled exposure to radiation with the devastating effects of high-level exposure, such as in nuclear explosions. However, modern reactors are equipped with multiple safety barriers to prevent radiation leaks, along with strict operational protocols.

Table 23: Major Nuclear Accidents

Accident	Year	Location	Main Consequences
Three Mile Island	1979	USA	Partial core meltdown, no direct fatalities
Chernobyl	1986	USSR	Reactor explosion, immediate deaths, and mass evacuation
Fukushima Daiichi	2011	Japan	Tsunami damage, evacuation, regional contamination

Source: Own elaboration based on the data presented in the Summary Table at the end of this chapter

The Challenge of Nuclear Waste

In addition to concerns about accidents, the management of nuclear waste is one of the most debated issues in society. The most common argument against nuclear energy is that its waste remains hazardous for thousands of years and that there is no definitive solution for its disposal.

However, this perception overlooks several essential facts:

- Over 90% of used nuclear fuel can be reused in closed fuel cycles, as is done in countries like France and Russia.

- There are effective and safe storage solutions, such as deep geological repositories.

- The volume of nuclear waste generated is significantly smaller than the waste produced by other industries. For comparison:

  - ○ A 1,000 MW nuclear power plant operating for an entire year generates only 30 tons of high-level radioactive waste.

  - ○ In contrast, a coal-fired plant of the same capacity produces over 300,000 tons of toxic ash per year.

What This Chapter Will Cover

This chapter aims to demystify the risks associated with nuclear safety and waste management. We will address:

- Technological advances that have made nuclear reactors increasingly safe.

- The different waste management strategies and viable solutions for disposal.

- Myths and truths about the dangers of nuclear waste.

- The future of nuclear safety and new technologies to minimize risks and enhance the sustainability of nuclear energy.

In this way, the reader will gain a perspective based on scientific data and modern engineering, free from the distortions commonly promoted by media alarmism and public misinformation.

Nuclear Plant Safety – Evolution and Technology

Nuclear safety is one of the most technologically advanced areas within the energy sector. From the first experimental reactors to today's Generation III+ and IV systems, significant advancements have been made to minimize operational risks and protect both the environment and the population.

The safety systems of a nuclear power plant can be categorized into three main pillars:

1. Physical containment barriers to prevent radiation leaks.

2. Redundant cooling systems to avoid reactor overheating.

3. Automatic shutdown procedures (safe shutdown) to mitigate operational failures.

In addition, new generations of reactors feature inherently safe designs that eliminate many of the vulnerabilities present in older models.

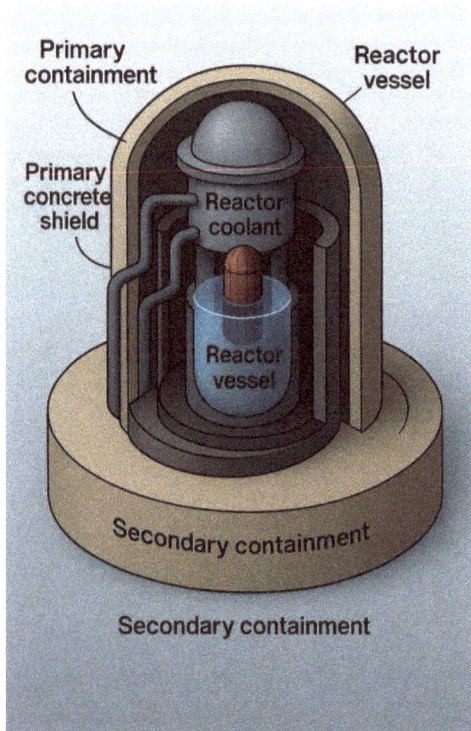

Illustration showing the detailed layout of the containment barriers in a modern nuclear reactor.

How Nuclear Safety Has Improved Over the Decades:

In the early nuclear era, safety concerns were limited due to a lack of incidents that revealed the real challenges of this technology. However, as reactors began to be used for commercial electricity generation, the first events emerged that highlighted the need for improvements.

Evolution of nuclear safety:

- **1950s–1960s**: First commercial reactors, based on military technology, with little emphasis on passive safety.

- **1970s–1980s**: Introduction of robust containment systems and emergency protocols after the Three Mile Island accident (1979).

- **1990s–2000s**: Improvements in cooling systems and automatic shutdown after Chernobyl (1986).

- **2000s–present**: Generation III+ reactors with passive safety, eliminating the need for human intervention to prevent core meltdowns.

Today, nuclear power plants feature multiple layers of safety, making catastrophic accidents like those of past decades highly unlikely.

Main Safety Systems in a Nuclear Reactor:

Modern nuclear reactors employ three primary safety layers to prevent radiation leaks and ensure plant integrity.

1. Containment Barriers

Modern nuclear reactors include three levels of barriers to prevent radioactive materials from escaping:

- **Nuclear fuel cladding** – A highly resistant ceramic layer of uranium oxide.

- **Reactor pressure vessel** – A high-strength steel metallic shell.

- **Containment building** – A reinforced concrete structure enclosing the entire facility.

2. Cooling Systems

Cooling the reactor core is critical for accident prevention. Current systems include:

- Primary and secondary circuits for heat transfer.

- Emergency diesel generators in case of external power failure.

- Passive cooling systems in modern reactors operating through natural convection without human intervention.

3. Automatic Shutdown Procedures

If any anomaly is detected, a nuclear reactor can shut down automatically within fractions of a second. Shutdown systems include:

- Automatic insertion of control rods to stop the chain reaction.

- Emergency ventilation to prevent pressure build-up in the system.

- Continuous monitoring via artificial intelligence, preventing failures before they occur.

Comparison Between Reactor Generations: What Has Changed?

The evolution of nuclear safety has not only occurred in reactor designs and technologies but also in how they are operated. Over the decades, there has been substantial investment in training for operators, engineers, and technicians, resulting in a highly qualified sector with increasingly strict safety standards.

Nuclear reactors are classified into Generations I, II, III, III+, and IV, each representing significant advancements in safety, efficiency, and emergency responsiveness. Below, we examine the technical evolution and its impact on operator training.

Table 24: Comparative Table of Reactor Generations

Generation	Period	Technical Characteristics	Operator Training and Safety Standards
Generation I	1950–1970	First commercial reactors, low efficiency, no automatic shutdown systems.	Manual operation, little regulation, rudimentary training.
Generation II	1970–2000	PWR and BWR become standard, first automatic shutdown	First mandatory certifications for operators, creation of

		systems, reinforced containment.	regulators such as the IAEA.
Generation III	2000–2020	Improved efficiency, more robust cooling systems, active safety measures.	Advanced simulations, rigorous training, and emergency simulators.
Generation III+	2020–present	Passive safety systems, core meltdown becomes nearly impossible.	More demanding international certifications, global standardization of best practices.
Generation IV	Future	Molten salt, thorium, self-sustaining concepts, closed fuel cycle.	AI-based operation, minimal human involvement in safety control.

Source: Own elaboration based on the data presented in the Summary Table at the end of this chapter

Comparative table between nuclear waste and fossil fuel waste. This helps contextualize the volume of waste generated, its environmental impact, and the available storage solutions.

Evolution of Technical Knowledge and Professional Training:

In the early years of nuclear energy, reactor operation was much more manual, requiring direct operator actions to control power output and safety conditions. There was no unified international regulation, and many early reactors were staffed by poorly trained teams, often composed of professionals transitioning from fields like electrical or mechanical engineering without specific training in nuclear physics.

Over time, there was a radical shift in the qualifications of nuclear power plant operators, accompanied by the implementation of strict training, education, and certification standards, which drastically reduced the risk of human error.

Key changes in nuclear operator training over the generations:

1. **Emergence of mandatory certification programs (1970s–1980s)**

After the Three Mile Island accident (1979), it became clear that human errors were a decisive factor in the severity of the incident.

In response, mandatory certification programs for reactor operators were introduced, including periodic exams and intensive training sessions.

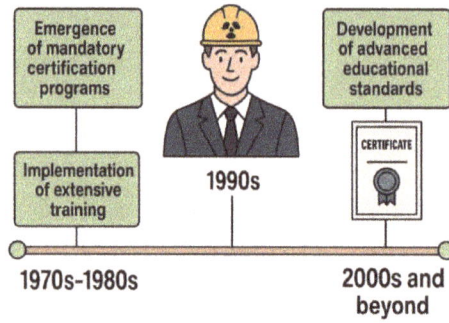

The illustration shows the timeline of the evolution of training, education, and certification of nuclear operators.

2. Introduction of Nuclear Simulators (1990s–2000s):

Inspired by the aviation industry, nuclear power plant simulators began to be used to train operators in severe failure scenarios.

This allowed teams to respond quickly to unexpected events without putting actual nuclear facilities at risk.

Illustration showing how nuclear operators are trained with an emergency simulator

Operators

Instructor

Illustration showing the setup of how nuclear operators are trained using emergency simulators.

3. **International Regulation and Global Standardization (2000s–Present)**

With the growth of nuclear energy, national and international regulatory agencies began defining standardized rules for professional training.

The International Atomic Energy Agency (IAEA), the U.S. Nuclear Regulatory Commission (NRC), and other entities made it mandatory to comply with continuous training and updated certifications.

4. Advanced Simulations and Artificial Intelligence in Reactor Operation (Near Future)

With the arrival of Generation IV reactors, training, and education will include virtual reality, artificial intelligence, and advanced simulations.

The role of the human operator will be minimized, as autonomous systems will continuously monitor reactor parameters and make automatic adjustments to optimize safety.

Chart 31: Evolution of Nuclear Reactor Safety and Efficiency

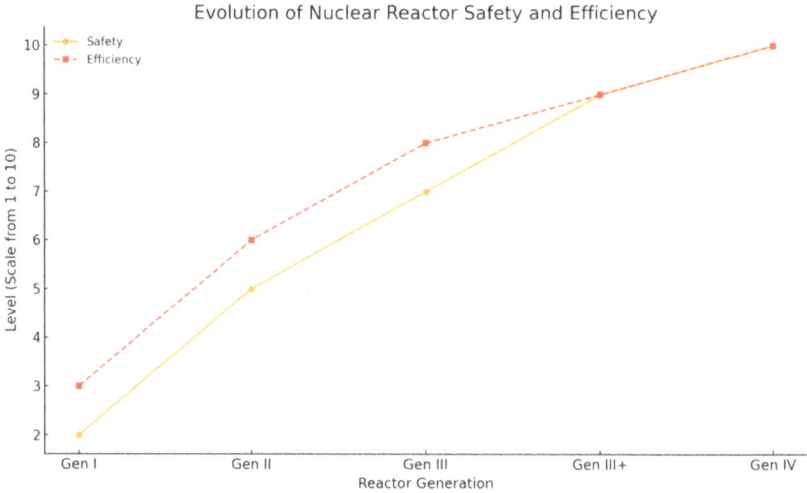

Source: Own elaboration based on the data presented in the Summary Table at the end of this chapter

The chart shows the evolution of safety and efficiency across nuclear reactor generations with the introduction of new technologies and safety procedures.

Successful Safety Case Studies:

The history of nuclear energy is filled with examples that demonstrate how improvements in safety have made modern reactors extremely reliable. Technological advancements, new materials, and rigorous operator training have drastically reduced the risk of accidents.

Although nuclear energy is often associated with disasters such as Chernobyl (1986) and Fukushima (2011), these incidents are exceptions, not the norm. Thousands of nuclear reactors have operated—and continue to operate—safely around the world, providing clean and stable energy for decades.

Here, we will examine three successful safety case studies that highlight the impact of enhanced nuclear technologies and protocols:

- The EPR reactor case in France – Advanced Safety.

- The Three Mile Island incident (1979) – An accident that proved the effectiveness of containment barriers.

- The impact of new reactor generations – How Generation III+ reactors could have prevented Fukushima.

The EPR Reactor Case in France – Advanced Safety

EPR (European Pressurized Reactor) units are among the most modern examples of nuclear safety. This Generation III+ design, developed by France's EDF and Germany's Siemens, represents the cutting edge of nuclear technology worldwide.

EPR Safety Measures

- **Double containment**: Two reinforced concrete buildings protect the reactor core.

- **Passive cooling system**: Operates through gravity and convection, eliminating the need for mechanical pumping.

- **Aircraft impact resistance**: Unlike older reactors, the EPR is designed to withstand large aircraft collisions.

- **Physical separation of emergency systems**: Ensures that multiple failures do not result in total loss of plant control.

Fun fact: The Flamanville 3 plant in France, equipped with an EPR reactor, was designed to be five times safer than conventional reactors.

Result: No recorded accidents in EPR reactors since their deployment! This is a clear example of how technological advances are making nuclear energy increasingly safer.

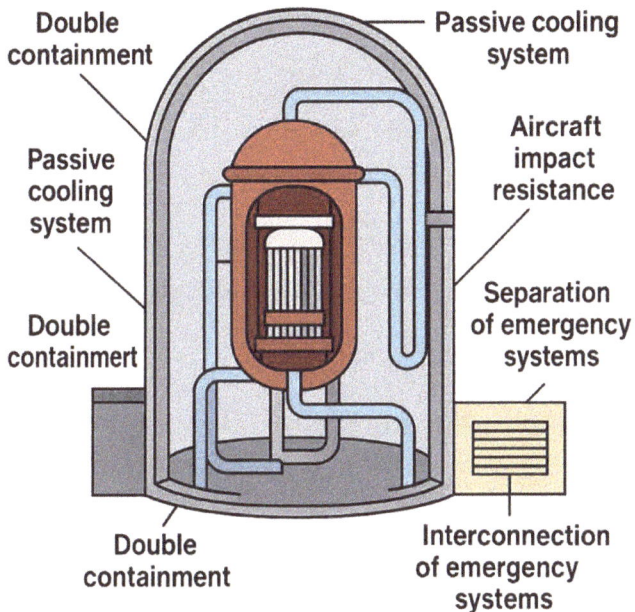

The illustration shows the detailed layout of the EPR reactor, highlighting its advanced safety layers.

The Three Mile Island Incident (1979) – An Accident That Proved the Effectiveness of Containment Barriers:

The Three Mile Island (TMI) accident in the United States in 1979 was one of the most studied events in nuclear history. Although classified as an 'accident,' this event actually served as a demonstration of the effectiveness of nuclear safety systems.

What happened at Three Mile Island?

- A safety valve failed, allowing coolant water to escape.

- Operators did not recognize the malfunction in time, leading to overheating of the core.

- Part of the nuclear fuel melted — **the first partial meltdown of a commercial reactor in the U.S**.

Why was this case considered a safety success?

- **No significant release of radiation occurred** — the containment barrier prevented radioactive material from escaping.

- **No deaths or public health impacts** were recorded.

- The accident led to a global overhaul of operator training protocols and certification standards.

THREE MILE ISLAND
CONTAINMENT BARRIER

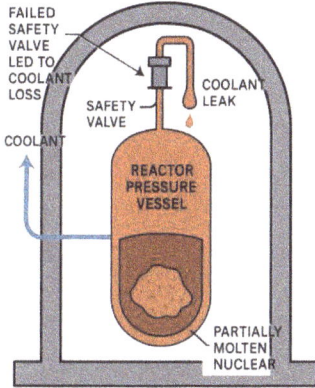

FAILED
SAFETY
VALVE
LED TO
COOLANT
LOSS
COOLANT
LEAK
SAFETY
VALVE
COOLANT
REACTOR
PRESSURE
VESSEL
PARTIALLY
MOLTEN
NUCLEAR

CONTAINMENT BARRIER PREVENTED
SIGNIFICANT RADIATION RELEASE

THREE MILE ISLAND
PREVENTED A NUCLEAR DISASTER

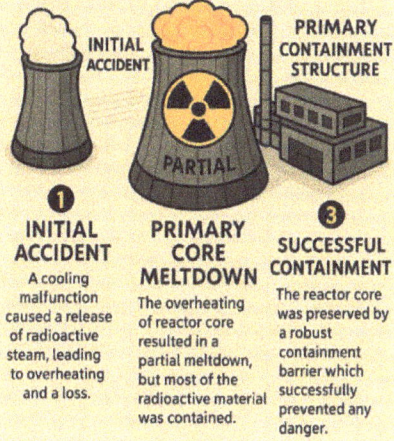

INITIAL
ACCIDENT
PRIMARY
CONTAINMENT
STRUCTURE
PARTIAL

1 INITIAL ACCIDENT
A cooling malfunction caused a release of radioactive steam, leading to overheating and a loss.

PRIMARY CORE MELTDOWN
The overheating of reactor core resulted in a partial meltdown, but most of the radioactive material was contained.

3 SUCCESSFUL CONTAINMENT
The reactor core was preserved by a robust containment barrier which successfully prevented any danger.

Infographic about the containment barrier at Three Mile Island, demonstrating how it prevented a nuclear disaster.

Conclusion:

Unlike Chernobyl, Three Mile Island proved that safety barriers work. If the same accident had occurred in an older reactor, it could have caused a tragedy. However, containment technologies showed they are capable of preventing major catastrophes.

189

Table 25: Nuclear Reactor Safety Barriers

Barrier	Protective Function
Fuel oxide film	Contains fission products within the fuel matrix
Metal cladding	Isolates fuel from the primary coolant circuit
Pressurized primary circuit	Prevents environmental exposure
Steel and concrete containment	Final barrier against radiation release

Source: Own elaboration based on the data presented in the Summary Table at the end of this chapter

The Impact of New Reactor Generations – How Generation III+ Reactors Could Have Prevented Fukushima

The Fukushima accident (2011) in Japan was one of the most impactful nuclear events of the 21st century. However, Generation III+ reactors are specifically designed to avoid this type of incident.

What happened at Fukushima?

- An earthquake followed by a tsunami destroyed the plant's power grid.

- The diesel generators failed, interrupting the cooling systems.

- Without cooling, the reactor cores overheated, and hydrogen explosions occurred.

If Fukushima had been equipped with Generation III+ reactors, the disaster likely would not have happened.

FUKUSHIMA vs. GENERATION III+

❶ **EARTHQUAKE AND TSUNAMI**
An earthquake and tsunami disbaled off-site power to the plant.

❷ **DIESEL GENERATORS FAILED**
Backup diesel generators were floded, stopping cooling systems.

❸ **CORE OVERHEATING**

PASSIVE SAFETY FEATURES

ENHANCED SAFETY FEATURES
Passive safety systems remove heat and maintain core cooling even without power

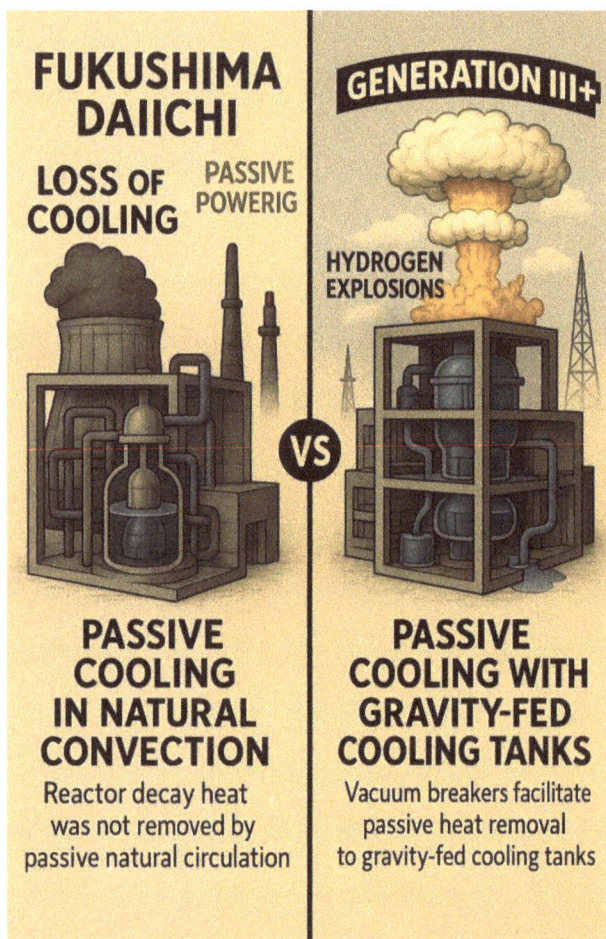

The illustration shows a comparative diagram illustrating the Fukushima failure versus the enhanced safety of a Generation III+ reactor.

Key Differences Between Fukushima and Modern Reactors:

Table 26: Differences Between Fukushima and Generation III+ Reactors

Fukushima (Generation II)	Generation III+ Reactors
Depended on external electricity for cooling	Passive cooling – operates without electricity.
Fragile containment structure	Reinforced structure against earthquakes and tsunamis
Hydrogen explosions due to a lack of ventilation	Hydrogen removal systems prevent explosive accumulation

Source: Own elaboration based on the data presented in the Summary Table at the end of this chapter

Fun fact: After Fukushima, all new nuclear power plants are required to include passive cooling systems.

Result: If Fukushima had been equipped with modern technology, the disaster would have been avoided. This demonstrates the tremendous evolution of nuclear safety over recent decades.

The cases presented clearly show that nuclear safety is not just theoretical but a proven reality:

- Modern reactors are designed to withstand extreme failures.

- Containment barriers truly work, as proven at Three Mile Island.

- Safety technology is constantly evolving, making nuclear energy increasingly reliable.

Today, Generation III+ and IV reactors make a nuclear disaster virtually impossible. With continued investment in research and innovation, nuclear energy is one of the safest and most efficient ways to generate electricity in the world.

The Practical Impact of Investment in Human Capital

The impacts of this major investment in professional training are clear. We can observe them by comparing two distinct cases:

Case of Human Error: Three Mile Island (1979)

- Operators manually shut down the cooling system due to misinterpretation of control panel signals.

- With proper training and education, the accident could have been avoided.

Case of Efficient Response: Fukushima Daiichi (2011)

- Even in the face of an extreme natural disaster (earthquake + tsunami), operators managed to prevent an even greater tragedy by delaying reactor collapse and implementing countermeasures.

- With the training improvements introduced since then, similar accidents may be mitigated or even completely avoided in the future.

The advancement of nuclear safety has not only come through reactor engineering but also through massive investment in the training of highly qualified professionals.

Today, a nuclear plant operator undergoes years of rigorous training, uses advanced simulators, and is subject to regular retraining to ensure readiness for any situation.

Generation III+ and IV reactors not only feature intrinsic safety but also rely on highly trained teams, ensuring that nuclear energy remains one of the safest and most reliable forms of electricity generation.

Nuclear Waste Management – The Real Challenge

The issue of nuclear waste is often cited as one of the greatest challenges of nuclear energy. However, contrary to alarmist narratives, radioactive waste has viable solutions for storage and recycling and is managed under high safety standards worldwide.

While fossil fuel waste is released directly into the atmosphere (CO_2, SO_2, NO_x), nuclear waste is confined and controlled from its production to final disposal. This means that, from an environmental perspective, nuclear waste is far more manageable than the waste from coal, oil, and gas.

In this section, we will explore the types of radioactive waste, current storage methods, and the real impact of decay time.

Types of Radioactive Waste

Nuclear waste is classified according to its level of radioactivity and the time required for its radiation to decay to a harmless level. The most common classification follows three main categories:

Table 27: Table of Nuclear Waste Types

Waste Type	Origin	Radioactivity Level	Management Method
Low-Level Waste	Hospital equipment, contaminated clothing, and plant tools	Low (decades of decay)	Temporary storage and safe disposal
Intermediate-Level Waste	Reactor structural components, resins, filters	Medium (decades to centuries of decay)	Concrete confinement or geological storage
High-Level Waste	Used nuclear fuel	High (centuries to millennia of decay)	Reprocessing or deep geological storage

Source: Own elaboration based on the data presented in the Summary Table at the end of this chapter

Low-Level Radioactivity

- Accounts for about 90% of the total volume of radioactive waste but has a very low level of radiation.

- Example: Medical equipment used in radiotherapy, gloves, and clothing used in nuclear power plants.

- This waste is stored for a few years until the radiation dissipates, after which it can be disposed of normally.

Intermediate-Level Radioactivity

- Makes up about 7% of the total waste volume.

- Includes parts from decommissioned reactors, resins, and contaminated sludge.

- This waste is encapsulated in concrete to prevent leaks and stored in secure locations.

High-Level Radioactivity

- Represents only 3% of the total volume but contains over 95% of the radioactivity of nuclear waste.

- The largest component is used fuel rods, which still contain significant energy potential.

- The main solutions for this type of waste are reprocessing or deep geological storage.

Fun fact: Many countries, such as France and Russia, reuse up to 95% of spent nuclear fuel, drastically reducing the amount of high-level radioactive waste.

Table 28: Classification of Nuclear Waste

Waste Type	Origin	Management Method
Low-Level Waste (LLW)	Clothing, tools, filters	Compaction and near-surface storage
Intermediate-Level Waste (ILW)	Reactor components, resins	Encapsulation and geological disposal
High-Level Waste (HLW)	Spent nuclear fuel	Cooling followed by deep geological storage

Source: Own elaboration based on the data presented in the Summary Table at the end of this chapter

Chart 32: Nuclear Fuel Utilization Rate

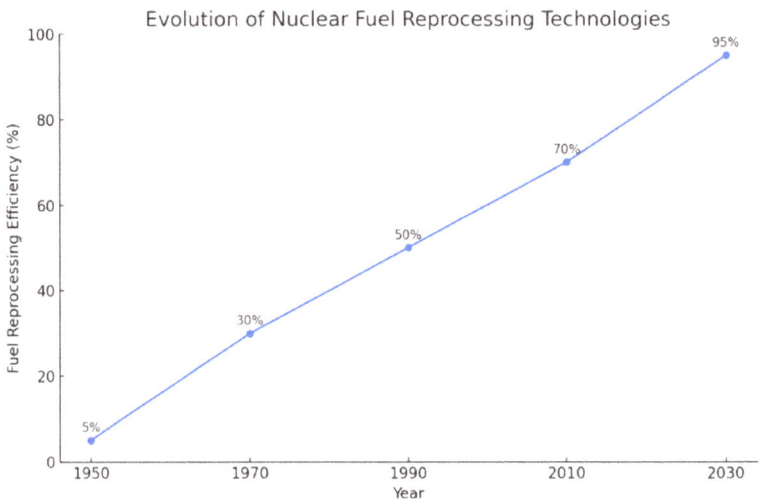

Evolution of Nuclear Fuel Reprocessing Technologies

Source: Own elaboration based on the data presented in the Summary Table at the end of this chapter

Chart showing the evolution of nuclear fuel recycling technologies, highlighting the increase in efficiency over the decades.

How Nuclear Waste Is Currently Stored

Nuclear waste management is one of the most regulated and controlled aspects of the nuclear energy industry. Unlike other industries that release pollutants directly into the environment (such as CO_2 from burning fossil fuels), nuclear waste is confined and managed with extreme safety, with no direct environmental impact.

Currently, there are three main methods of nuclear waste storage used at different stages of the radioactive waste lifecycle:

1. Cooling Pools – The First Stage of Storage

When nuclear fuel is removed from the reactor, it still has high radioactivity and residual heat. To cool it and reduce its initial radioactivity, it is temporarily stored in cooling pools within the nuclear plant itself.

How do cooling pools work?

- They are large tanks filled with demineralized water, built with reinforced concrete structures.

- The water absorbs radiation and dissipates the fuel's heat.

- After 5 to 10 years, the material can be removed from the pool and either dry stored or reprocessed.

Fun fact: Even in the worst-case scenario of a water leak, radiation would still be contained by the pool's structure, avoiding any environmental risk.

2. Dry Casks – The Medium-Term Solution

After the used nuclear fuel has cooled in pools, it can be transferred to sealed steel and concrete containers known as dry casks.

Advantages of dry casks:

- They do not require electricity for cooling and operate passively.

- They are highly resistant to impact, earthquakes, fire, and explosions.

- They can safely store used nuclear fuel for over 100 years.

Example: The United States has used dry casks since the 1980s and has never experienced a radiation leak or accident associated with this storage method.

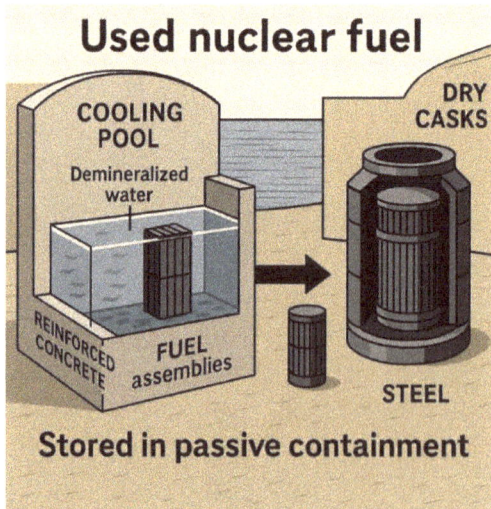

The illustration shows how used nuclear fuel is stored in cooling pools and dry casks.

3. Deep Geological Storage – The Definitive Solution

High-level radioactive waste, which remains hazardous for centuries or millennia, requires a permanent and secure repository. The most advanced solution is deep geological storage, which involves burying waste hundreds of meters underground in stable rock formations.

How does geological storage work?

- Waste is placed in metal containers coated with copper.

- These containers are buried in tunnels excavated into extremely old and stable rock.

- The geological barrier prevents any radiation from migrating to the surface.

The world's most advanced project of this kind is the Onkalo repository in Finland.

Case Study: Onkalo – The World's First Geological Repository

Onkalo, in Finland, is the first nuclear waste repository in the world to begin operations. This project was developed to safely store high-level radioactive waste for 100,000 years.

Why did Finland choose Onkalo?

- Finland has one of the most stable geological formations on the planet, composed of granite over 1.8 billion years old.

- The country produces 30% of its electricity from nuclear energy and needed a safe, permanent solution for the waste.

- Onkalo was designed to be completely self-sufficient, requiring no maintenance after sealing.

How does Onkalo work?

- Location – The repository is located 450 meters below the surface, inside a rocky massif.

- Protection layers – The waste is stored in copper capsules, which are surrounded by bentonite clay to prevent water infiltration.

- Final structure – Once the repository is full (expected by 2120), the tunnels will be permanently sealed.

Key Benefits of the Onkalo Repository

- Absolute safety – Even in the event of earthquakes or geological activity, the radiation will remain isolated.

- Zero environmental impact – The system is designed to prevent any leakage into the environment.

- Model for the future – Other countries, such as Sweden and France, are planning similar repositories.

Fun fact: Studies indicate that even if humanity were to disappear, the waste stored in Onkalo would remain isolated and safe without any need for human intervention.

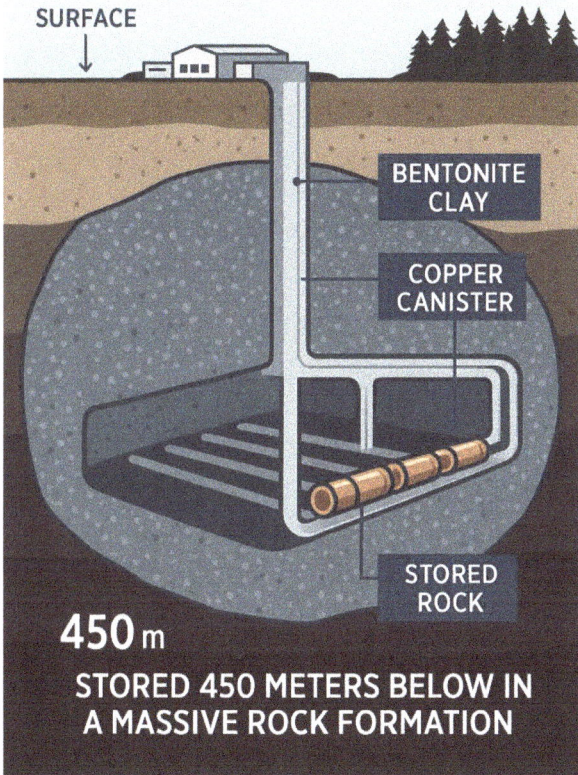

Infographic of the Onkalo Geological Repository, showing its underground structure and protective layers.

Conclusion:

Contrary to what many believe, nuclear waste is not a problem without a solution. On the contrary, it has one of the most secure and rigorously managed chains in the world.

- Cooling pools ensure immediate safety for waste freshly removed from the reactor.

- Dry casks offer a highly secure medium-term solution.

- Deep geological storage, like Onkalo, represents a definitive and risk-free solution.

COMPARISON OF NUCLEAR WASTE STORAGE METHODS

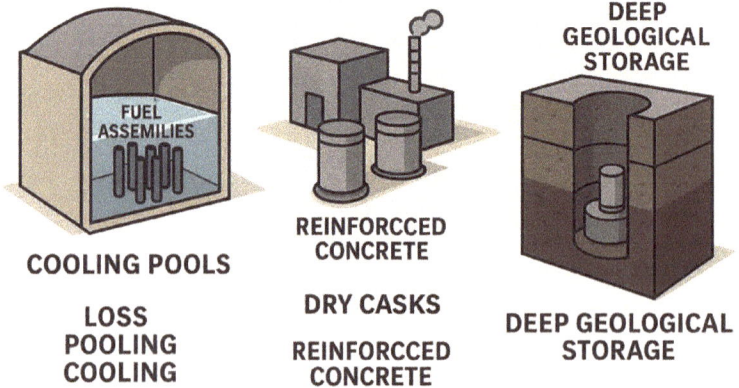

DEEP GEOLOGICAL STORAGE

FUEL ASSEMILIES

REINFORCCED CONCRETE

COOLING POOLS

LOSS POOLING COOLING

DRY CASKS

REINFORCCED CONCRETE

DEEP GEOLOGICAL STORAGE

The illustration shows a comparative diagram of different nuclear waste storage methods, including cooling pools, dry casks, and deep geological storage.

Nuclear energy is the only form of electricity generation that takes full responsibility for 100% of its waste, ensuring that no environmental impact occurs now or in the future.

Decay Time and the Myths of "Eternal Danger" in Nuclear Waste

One of the greatest misconceptions about nuclear waste is the belief that it remains dangerously radioactive for hundreds of thousands of years with no viable solution. Although some isotopes have long half-lives, the reality is that most nuclear waste loses 99% of its radioactivity within just a few hundred years.

In addition, many countries use reprocessing to reduce the volume of high-level radioactive waste. France, for example, recycles more than 80% of its used nuclear fuel, recovering valuable materials and minimizing the environmental impact of the remaining waste.

In this section, we will explore how long nuclear waste actually takes to decay and how nuclear fuel recycling can drastically reduce the scope of this issue.

1. Decay Time of Nuclear Waste

The radioactivity of nuclear waste decreases over time as radioactive isotopes decay into stable and harmless elements. The pace of this process is measured by the half-life—the amount of time it takes for half of the radioactive atoms to decay.

Table 29: Table Showing the Decay Times of Nuclear Waste

Material	Half-life (time to lose half of the radioactivity)	Estimated Time to Safe Level
Iodine-131	8 days	3 months
Cesium-137	30 years	300 years
Strontium-90	29 years	300 years
Plutonium-239	24.000 years	240.000 years

Source: Own elaboration based on the data presented in the Summary Table at the end of this chapter

Facts About Radioactivity Decay in Nuclear Waste

Fun fact: About 90% of the radioactivity in used nuclear fuel disappears within the first 300 years.

Another important fact: Elements like Iodine-131, which are highly radioactive, decay almost completely in just a few months.

This means that most nuclear waste is not dangerous for hundreds of thousands of years, as many believe. Only a small fraction of the waste requires long-term storage—and that fraction can be reduced through reprocessing technologies.

DECLINE OF RADIOACTIVE WASTE OVER TIME

Iodine-131	**8** days
Cesium-137	**30** years
Plutonium-239	**24,000** years
Technetium-99	**210,000** years

| Days | Years | Centuries | Millennia |

TIME

Illustration showing the decay time of major radioactive elements.

2. How France Recycles Over 80% of Its Nuclear Fuel

France is a global leader in nuclear fuel recycling, drastically reducing the amount of high-level radioactive waste and increasing the energy efficiency of its reactors. The country operates a closed fuel cycle, allowing valuable materials to be reused in new nuclear reactions.

How does French reprocessing work?

- Material Separation
  - Used nuclear fuel is sent to the La Hague reprocessing facility, one of the largest in the world.

- There, uranium and plutonium are separated from the highly radioactive waste.

- Reuse of Uranium and Plutonium

  - The separated uranium can be reconverted into new fuel for reactors.

  - The extracted plutonium is used to produce MOX (Mixed Oxide Fuel), a type of fuel that can be reused in nuclear reactors.

- Reduction of High-Level Waste

  - Only 3% of the initial fuel becomes true waste with no reuse potential.

  - This waste is vitrified—mixed with molten glass and safely stored in geological repositories.

What does France do with the recycled fuel?

- Uses MOX (Mixed Oxide Fuel) – A fuel made from recycled uranium and plutonium, used in about 22 French nuclear reactors.

- Avoids excessive uranium mining – Reducing the need for extracting new natural resources.

- Reduces long-term waste – What would normally require 240,000 years of storage can be reduced to just a few centuries through reprocessing.

Fun fact: Thanks to reprocessing, France produces less than 1 kg of high-level radioactive waste per capita per year — much less than most countries that use nuclear energy.

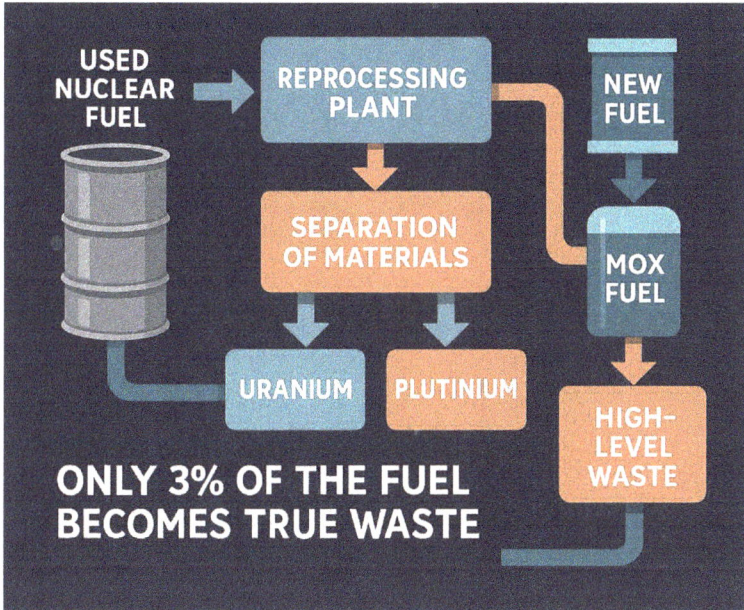

Flowchart illustrating the nuclear fuel reprocessing cycle in France, highlighting how over 80% of the fuel is reused.

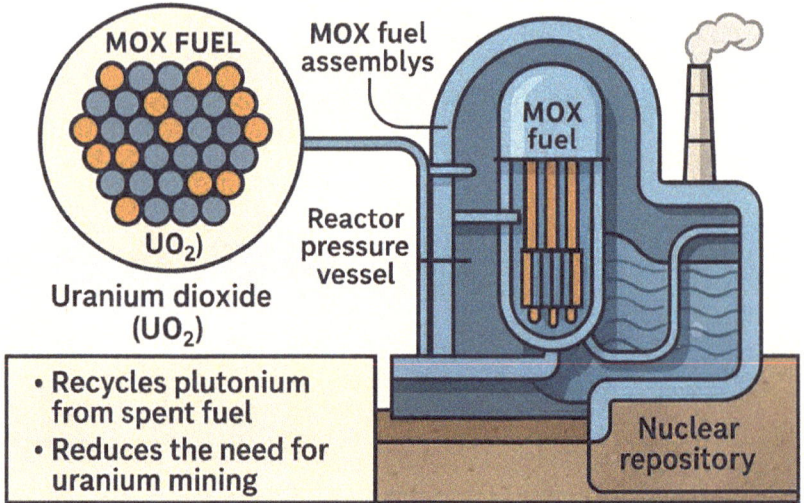

MOX FUEL IN NUCLEAR REACTOR

MOX FUEL

UO_2)

Uranium dioxide (UO_2)

MOX fuel assemblys

Reactor pressure vessel

MOX fuel

Nuclear repository

- Recycles plutonium from spent fuel
- Reduces the need for uranium mining

A diagram illustrates how MOX (Mixed Oxide Fuel) works in nuclear reactors, highlighting its composition and benefits.

3. The Future of Waste Management – New Recycling Technologies

Advancements in next-generation nuclear reactors will allow nearly 100% of nuclear fuel to be reused, making the resulting waste even smaller. Some promising technologies include:

Fast Reactors

- Use plutonium and depleted uranium as fuel, closing the nuclear fuel cycle.

- Are extremely efficient and can reduce nuclear waste by up to 90%.

Molten Salt Reactors

- Enable more efficient and safer nuclear reactions.

- Can consume older nuclear waste, turning it into new energy sources.

Nuclear Transmutation

- An experimental technology that can convert long-lived isotopes into shorter-lived elements.

- This could eliminate the need for long-term storage.

Example: Japan and the European Union are investing billions in nuclear transmutation research, aiming to minimize long-term waste.

Conclusion – Nuclear Waste Is a Smaller Problem Than Commonly Believed

Contrary to popular belief, nuclear waste is not an unsolvable problem. On the contrary, there are effective technologies and strategies for managing it:

- Most of the waste's radioactivity disappears within a few centuries.

- Reprocessing can drastically reduce the volume of high-level waste.

- Countries like France already reuse 80% of nuclear fuel, minimizing environmental impact.

- New technologies may transform nuclear waste into new energy sources.

When properly managed, nuclear waste does not pose a significant risk to the environment. With innovation and strong management policies, the nuclear sector can become even more sustainable in the future.

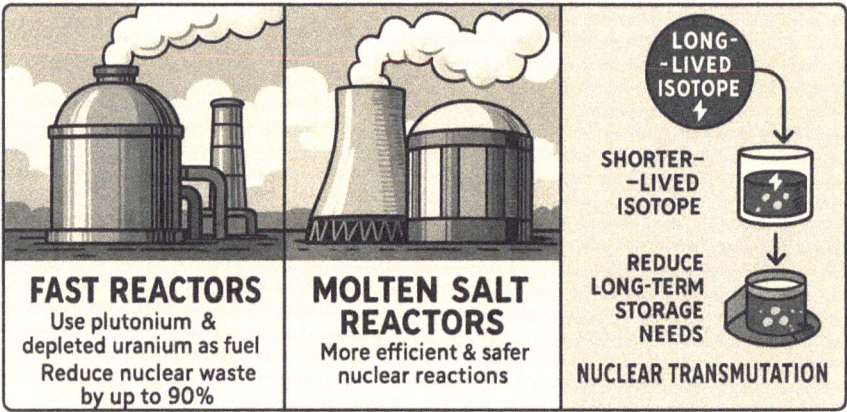

NEW NUCLEAR RECYCLING TECHNOLOGIES
Promising technologies that can drastically reduce radioactive waste

FAST REACTORS
Use plutonium & depleted uranium as fuel
Reduce nuclear waste by up to 90%

MOLTEN SALT REACTORS
More efficient & safer nuclear reactions

LONG--LIVED ISOTOPE

SHORTER--LIVED ISOTOPE

REDUCE LONG-TERM STORAGE NEEDS

NUCLEAR TRANSMUTATION

Infographics illustrate new nuclear recycling technologies, including fast reactors, molten salt reactors, and nuclear transmutation, which can drastically reduce radioactive waste.

Myths About the Danger of Nuclear Waste

Nuclear waste management is one of the most frequently criticized topics by opponents of nuclear energy. Waste is often portrayed as 'eternal garbage,' unsolvable and extremely dangerous. However, this view overlooks three key points:

- The amount of nuclear waste generated is extremely small compared to other industries.

- Nuclear waste is stored and controlled, whereas chemical and fossil fuel waste is released into the environment.

- The radioactivity of nuclear waste decreases over time, while chemical pollutants and heavy metals remain toxic forever.

Let's explore these points in depth, debunking the main claims about nuclear waste.

Comparison Between Nuclear Waste and Other Industries

Nuclear energy generates waste, but so does every form of energy generation. The central issue is not simply whether a sector generates waste but how that waste is treated and what real impact it has on the environment and human health.

Let's compare nuclear waste with waste from other industries:

Table 30: Comparison Between Nuclear Waste and Other Industrial Waste

Waste Type	Origin	Annual Quantity Produced	Environmental Impact	Management
Nuclear Waste	Nuclear power plants	About 30 tons per 1GW reactor/year	No direct environmental impact (safely stored)	Reprocessing and geological storage
Coal Ash	Thermal power plants	Millions of tons	Contain toxic heavy metals (mercury,	Disposed in landfills, often leak

			arsenic) contaminating soil and water	into the environment
Chemical Waste	Petrochemical and pharmaceutical industries	Billions of liters of toxic effluents	Pollution of rivers, aquifers, and population poisoning	Partial treatment and ongoing discharge
Electronic Waste	Tech trash (batteries, boards, etc.)	Millions of tons	Contain lead, cadmium, and mercury – highly toxic	Limited recycling, common improper disposal

Source: Own elaboration based on the data presented in the Summary Table at the end of this chapter

Why Nuclear Waste Is Less Problematic and the Reality Behind Industrial Risks:

- Smaller volume – Nuclear waste is produced in minute quantities compared to other types of industrial waste.

- Safe storage – Unlike chemical and coal waste, nuclear waste is not released into the environment.

- Radioactive decay – While chemical pollutants and heavy metals remain permanently toxic, nuclear waste becomes less dangerous over time.

Fun fact: A nuclear power plant supplies electricity to millions of people and generates only 30 tons of high-level radioactive waste per year. A coal plant of the same capacity generates

300,000 tons of toxic ash annually, with part of it released into the atmosphere.

The Real Risk of Radioactive Waste Compared to Public Perception

Public opinion often overestimates the risks of nuclear waste while underestimating the risks of other waste. This is largely due to decades of amplified fear of radioactivity, while invisible industrial pollutants are widely accepted without question.

Let's break down some common myths:

Myth 1: Nuclear waste is dumped into the environment

FALSE. No industry controls its waste as strictly as the nuclear sector.

100% of nuclear waste is stored in protected and regulated facilities. Meanwhile, chemical waste is often dumped into rivers, and coal ash is released into the air.

Myth 2: Nuclear waste is dangerous forever

FALSE. Most of the radioactivity disappears within a few hundred years.

Elements like Cesium-137 and Strontium-90 (the most hazardous in the short term) lose 99% of their radiation in 300 years. Only a small fraction of waste needs long-term storage.

Myth 3: Nuclear waste is more dangerous than chemical waste

FALSE. Substances like mercury, cadmium, and industrial pesticides never lose their toxicity.

Contamination by mercury or dioxins is irreversible, while radioactive contamination naturally dissipates over time.

Fun fact: The Bhopal chemical disaster in India (1984) killed over 15,000 people. No nuclear waste leak or accident has ever caused anything remotely similar.

The Bhopal Chemical Disaster (India, 1984) – The Real Danger of Industrial Waste

On the night of December 2–3, 1984, the city of Bhopal, India, suffered the worst chemical disaster in history. A severe leak of the toxic gas methyl isocyanate (MIC) at the Union Carbide India Limited (UCIL) plant resulted in the death of over 15,000 people and hundreds of thousands of severe poisoning cases, many with permanent consequences.

This catastrophic event is often overlooked in industrial safety discussions, yet its scale far exceeds any nuclear disaster to date.

1. What Happened in Bhopal?

The UCIL pesticide plant stored methyl isocyanate (MIC), a highly toxic gas used in pesticide production. On the night of the accident, water entered the MIC storage tank, triggering an uncontrollable chemical reaction.

Disaster timeline:

- 10:30 PM – Water infiltrates MIC storage tanks.

- 11:00 PM – The heat from the chemical reaction increases internal tank pressure.

- 12:30 AM – The safety valve fails, releasing a massive cloud of toxic gas over the city.

- 1:00 AM – Thousands of people begin dying instantly after inhaling the gas.

- By dawn – Over 5,000 were confirmed dead and around 600,000 poisoned.

The safety systems failed, and the population was not evacuated in time, resulting in devastating consequences.

2. Impact of the Disaster – A Human and Environmental Catastrophe

Health Impact

- 15,000 to 25,000 direct and indirect deaths in the following years.

- Over 600,000 people were poisoned, many with permanent health issues.

- Thousands of cases of blindness and chronic respiratory problems.

The MIC gas attacked lungs, eyes, and internal tissues, causing agonizing deaths from suffocation and internal burns.

Environmental Impact

- Soil and water contaminated for decades – the factory continued leaking chemicals for years.

- Groundwater contamination caused severe illnesses in the local population.

- The area around the factory remains polluted to this day.

Unlike a nuclear disaster, where radiation dissipates over time, chemical pollutants remain in the environment and continue to affect people nearly 40 years later.

Comparison with Nuclear Accidents

The Bhopal disaster was far more devastating than any nuclear event in terms of casualties, environmental impact, and corporate negligence.

Table 31: Comparison Between the Bhopal and Chernobyl Accidents

Criterion	Bhopal (1984)	Chernobyl (1986)
Cause	Industrial failure and lack of chemical safety	Nuclear reactor explosion
Immediate deaths	5,000 to 10,000	31
Deaths over the years	15,000+	4,000–10,000 (WHO estimate)
People affected	600,000+ intoxicated	200,000 evacuated
Contaminated area	Soil and water are permanently polluted	Radiation has decreased over the years
Current impact	Chemical waste still pollutes the region	Radiation has dropped by over 90%
Accountability	The company paid minimal compensation	International nuclear safety measures reinforced

Source: Own elaboration based on the data presented in the Summary Table at the end of this chapter

218

Important fact: Unlike nuclear energy, where accidents like Chernobyl led to massive reforms in safety protocols, the chemical industry still operates, in most cases, with high risks, experiencing multiple leaks and toxic pollution incidents every year.

Chart 33: Nuclear Fuel Utilization Rate

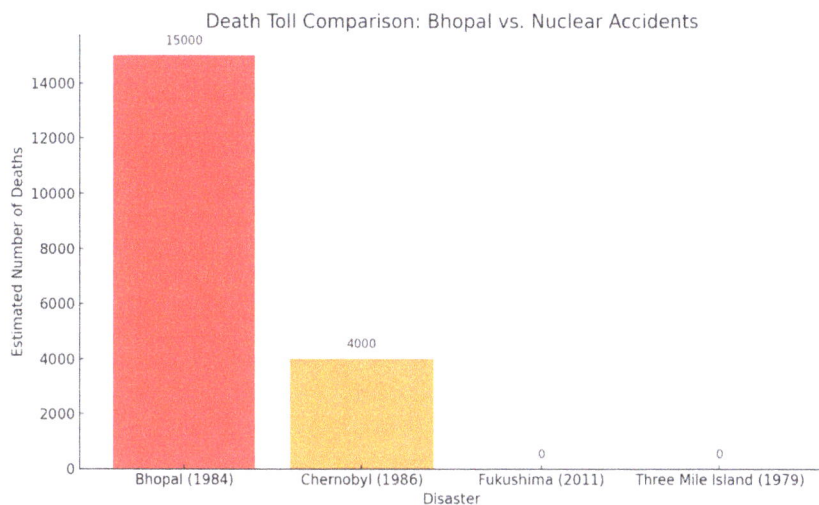

Death Toll Comparison: Bhopal vs. Nuclear Accidents

Source: Own elaboration based on the data presented in the Summary Table at the end of this chapter

Why Is There No Global Fear of the Chemical Industry?

Despite Bhopal being far worse than any nuclear accident, the chemical industry does not face the same pressure and scrutiny as the nuclear sector. This happens because:

- Lack of public knowledge – Most people don't understand chemical risks the way they understand the danger of radiation.

- Economic interests – Chemical and oil industries are powerful and influential.

- Lack of strict regulation – Unlike the nuclear sector, control over the chemical industry is much looser.

Curious fact: A single chemical disaster killed more people than all nuclear accidents combined, yet the fear remains focused on nuclear energy.

BHOPAL vs CHERNOBYL
GAS LEAK   NUCLEAR ACCIDENT

15,000+ DEATHS   4,000 DEATHS

LASTING SOIL AND WATER CONTAMINIATION   RADIATION DISPERSAL OVER TIME

HUMAN IMPACT   IMPACT

Conclusion – Bhopal and the Hypocrisy in Risk Perception

The Bhopal disaster should have been a global wake-up call about the risks of the chemical industry, but instead, it continues to be overlooked.

The Bhopal accident killed over 15,000 people, while the worst nuclear accident (Chernobyl) had a significantly smaller impact.

Chemical contamination in Bhopal continues to this day, whereas radiation in Chernobyl has already decayed by more than 90%.

Nuclear safety regulations were strengthened after Chernobyl, but the chemical industry continues to operate with high risks.

What does Bhopal teach us? The real danger to the environment and public health does not come from nuclear waste but from industrial and chemical waste, which is dumped uncontrollably around the world.

TIMELINE OF THE BHOPAL DISASTER

10:30 PM

Water leaks into a methyl isocyanate storage tank, triggering an exothermic reaction

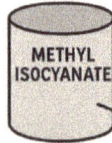

11:00 PM

Temperature and pressure inside the tank begin to rise

12:30 AM

The safety valve fails, and a large cloud of toxic gas escapes into the air

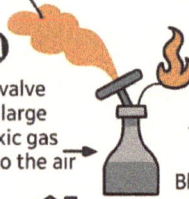

1:00 AM

Toxic gas reaches nearby city of Bhopal, thousands of citizens start to die

By morning, at least 3,000 people have died from exposure

By morning, at least 3,000 people have died from exposure to the gas

Timeline of the Bhopal Disaster, illustrating the events that led to the catastrophe.

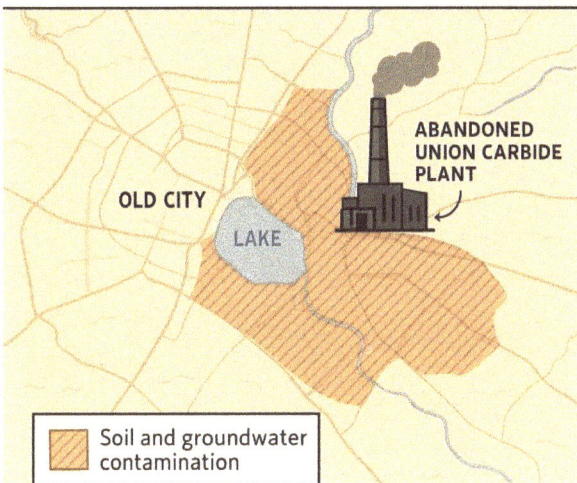

MAP OF CHEMICAL CONTAMINATION IN BHOPAL

Areas still affected by the industrial disaster to this day

ABANDONED UNION CARBIDE PLANT

OLD CITY

LAKE

Soil and groundwater contamination

Illustration showing the chemical contamination map in Bhopal, highlighting the areas still affected by the disaster to this day.

Real Contamination Cases vs. Disproportionate Alarmism:

Nuclear waste has never caused a global environmental disaster, while chemical pollutants and oil spills have caused irreversible damage.

Table 32: Comparison of Environmental Disasters

Case	What happened?	Real impact	Problem management
Chernobyl (1986)	The reactor explosion released radioactive material	Local area affected, 30 km evacuation zone	Radiation has dropped by over 90% since 1986
Coal ash spill (Kingston, USA, 2008)	Reservoir rupture released millions of tons of toxic ash	Rivers and lands are permanently contaminated	Irreversible environmental impact
BP – Gulf of Mexico (2010)	The platform explosion released millions of barrels of oil into the ocean	Species extinction, long-lasting environmental damage	Partial remediation, irreversible damage

Source: Own elaboration based on the data presented in the Summary Table at the end of this chapter

Conclusion: No nuclear accident has caused an environmental impact comparable to chemical and oil disasters.

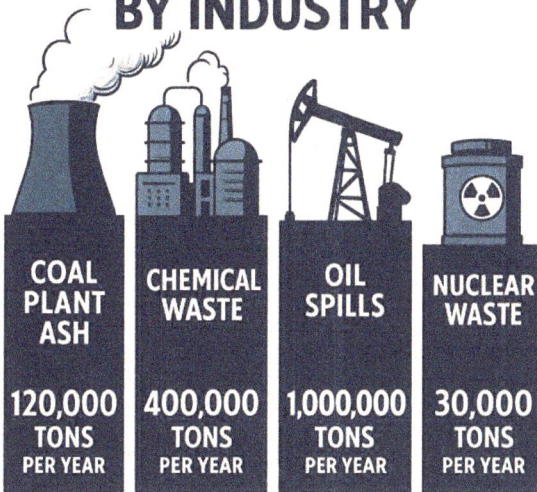

WASTE GENERATION BY INDUSTRY

COAL PLANT ASH	CHEMICAL WASTE	OIL SPILLS	NUCLEAR WASTE
120,000 TONS PER YEAR	400,000 TONS PER YEAR	1,000,000 TONS PER YEAR	30,000 TONS PER YEAR

Comparative infographic illustrating the amount of waste generated by different industries, highlighting the small quantity of nuclear waste compared to other types of pollutants.

The BP Disaster in the Gulf of America (2010) – The Real Danger of the Oil Industry:

On April 20, 2010, the Deepwater Horizon drilling platform, operated by British Petroleum (BP) in the Gulf of America, suffered a catastrophic explosion, causing the largest oil spill in U.S. history.

The disaster resulted in 11 immediate deaths, hundreds of injuries, destruction of marine ecosystems, and contamination that persists to this day.

This event is crucial for comparing the real risks of the oil industry with those of nuclear energy, as no nuclear power plant has ever caused environmental damage comparable to this spill[2].

1. What Happened in the BP Disaster?

Deepwater Horizon was a drilling platform operating in deep waters, exploring an oil well in the Macondo field, over 1,500 meters deep.

On the night of April 20, 2010, a catastrophic failure in the well's cementing allowed high-pressure natural gas to escape uncontrollably, reach the platform, and ignite.

Disaster timeline:

April 20, 9:49 PM – Natural gas begins leaking from the Macondo well into the platform's piping.

April 20, 9:56 PM – The gas reaches the deck of Deepwater Horizon and ignites.

April 20, 10:00 PM – A massive explosion destroys the platform, killing 11 workers and injuring dozens.

April 22 – The platform sinks, completely rupturing the well and triggering an uncontrolled oil spill.

---

[2] The author, having worked in the oil industry for over thirty years, feels ashamed by the accident and all its consequences. The oil industry has always upheld the highest standards of quality and safety, and witnessing an accident of this magnitude—clearly caused by human negligence both at the company's management level and among its operational leaders— evokes a deep sense of repulsion and indignation. Justice has done its part, appropriately convicting the company and its managers for gross negligence.

July 2010 – After 87 days, BP finally manages to seal the well. During that period, more than 4.9 million barrels of oil were released into the ocean.

This massive oil spill became the largest environmental disaster ever caused by the oil industry in the United States.

TIMELINE OF THE BP OIL SPILL
IN THE GULF OF AMERICA

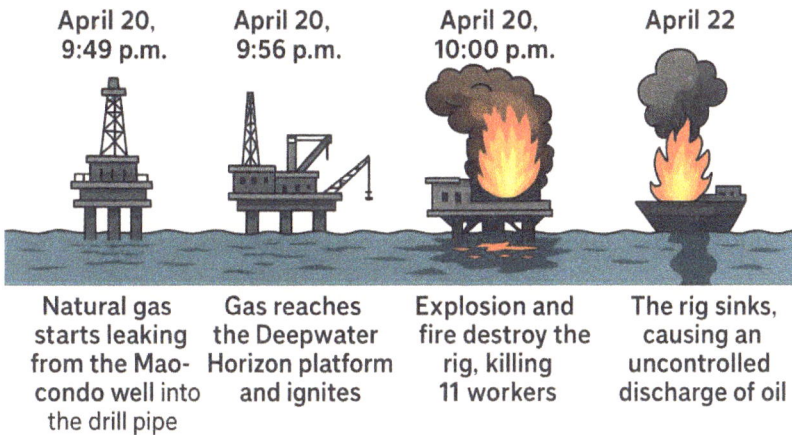

April 20, 9:49 p.m.	April 20, 9:56 p.m.	April 20, 10:00 p.m.	April 22
Natural gas starts leaking from the Mao-condo well into the drill pipe	Gas reaches the Deepwater Horizon platform and ignites	Explosion and fire destroy the rig, killing 11 workers	The rig sinks, causing an uncontrolled discharge of oil

Timeline of the BP Disaster in the Gulf of Mexico, illustrating the events that led to the oil spill.

2. Environmental Impact of the BP Oil Spill

While nuclear disasters like Chernobyl and Fukushima caused localized damage, the BP spill had a global impact, affecting the ocean, marine life, and the fishing economy for decades.

Impact on Marine Life

- Over 100,000 sea turtles were killed.

- At least 1,400 dolphins and whales died as a result of the contamination.

- The fishing industry lost billions of dollars due to the destruction of marine ecosystems.

Impact on the Economy

- Over 400,000 jobs affected the fishing and tourism industries.

- BP paid more than $65 billion in compensation, but the environmental damage is irreparable.

Impact on the Environment

- The oil slick covered an area equivalent to the state of New York.

- More than 4.9 million barrels of oil spilled, contaminating 1,300 km of coastline.

- Plankton, coral reefs, and shellfish were devastated, permanently altering the region's food chain.

Important fact: Radiation from Chernobyl and Fukushima has naturally declined over the years, while BP's oil continues to contaminate the ocean to this day.

Map showing the extent of the oil spill in the Gulf of Mexico, highlighting the areas affected by the contamination.

3. Comparison with Nuclear Accidents

Despite the massive destruction caused by the BP disaster, the public still fears nuclear energy more than oil, even though oil-related disasters are recurrent and cause permanent environmental damage.

Table 33: Comparison Between the BP and Chernobyl Accidents

Criterion	BP Gulf of Mexico (2010)	Chernobyl (1986)
Cause	Failure in oil well cementing	Nuclear reactor explosion
Immediate deaths	11	31
Deaths over the years	Est. 11,000 due to contamination and economic impact	4,000–10,000 (WHO estimate)
Affected area	1,300 km of coastline, open ocean	30 km around the reactor
Environmental impact	Oil spreads in the ocean, causing permanent contamination	Radiation has decreased over the years
Recovery	Environmental impact is still present after 14 years	Radiation has dropped by more than 90% since 1986
Accountability	BP paid compensation, but environmental damage persists	Global nuclear safety protocols reinforced

Source: Own elaboration based on the data presented in the Summary Table at the end of this chapter

Conclusion: The BP disaster was far more harmful to the environment and the economy than any modern nuclear accident.

Chart 34: Comparison of Environmental Impact – BP Disaster vs. Nuclear Energy

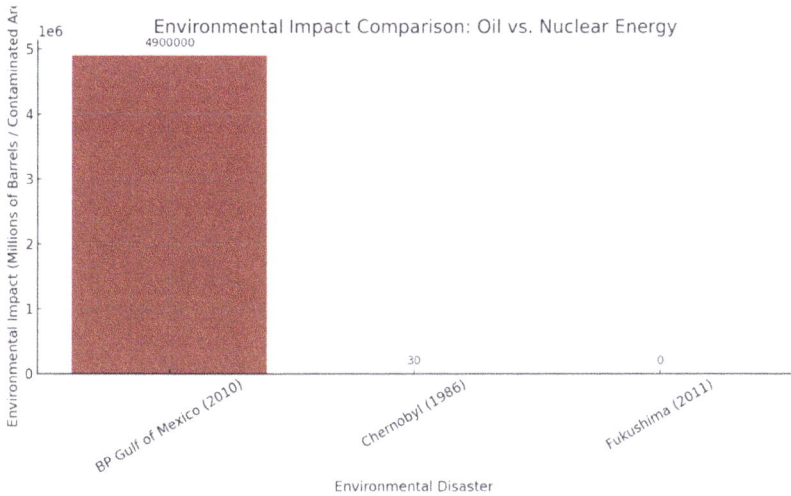

Environmental Impact Comparison: Oil vs. Nuclear Energy

Source: Own elaboration based on the data presented in the Summary Table at the end of this chapter

Chart comparing the environmental impact of oil vs. nuclear energy, showing the disparity between disasters.

Note: To enable a clear comparison, the Chernobyl accident was converted into an environmental impact equivalent to barrels of spilled oil. Thus, the Chernobyl accident had an environmental impact equivalent to a spill of 30,000 barrels of oil.

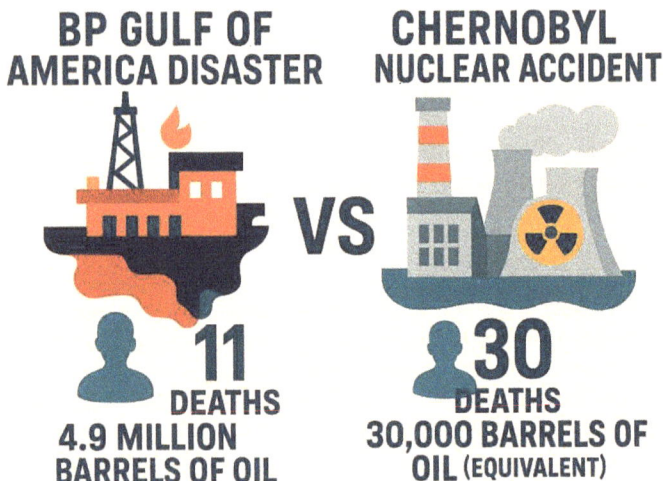

Visual comparison between the BP disaster in the Gulf of America and Chernobyl, highlighting the human and environmental impact of each event.

4. Why Is Oil Less Feared Than Nuclear Energy?

The oil industry moves trillions of dollars and has direct influence over global politics. Thus, oil disasters—though rare—have devastating consequences but are quickly forgotten. Much like nuclear energy, oil-related accidents often lead to improvements in industry practices and procedures. However, nuclear energy continues to suffer from a long-lasting stigma.

The oil lobby is powerful – Oil companies fund political campaigns and control media narratives.

Oil is part of daily life – As we depend on fossil fuels, the public tends to overlook their environmental impacts.

The fear of radiation outweighs the fear of chemical pollution – Misinformation about nuclear waste fuels paranoia, while real environmental disasters are downplayed.

The BP disaster was far more destructive and long-lasting than any nuclear accident, yet the fear of nuclear energy persists while oil continues to be widely consumed.

The BP oil spill caused permanent contamination of the ocean, whereas the radiation from Chernobyl has significantly decayed.

BP paid billions in compensation, but the environmental damage remains irreparable.

Oil spills continue to occur regularly but are quickly forgotten.

The true danger does not lie in nuclear energy but in our dependence on fossil fuels, which cause massive environmental disasters, destruction of ecosystems, and climate change.

Illustration showing the visual comparison between environmental disasters (nuclear vs. chemical vs. oil), highlighting differences in impact, environmental damage, and recovery.

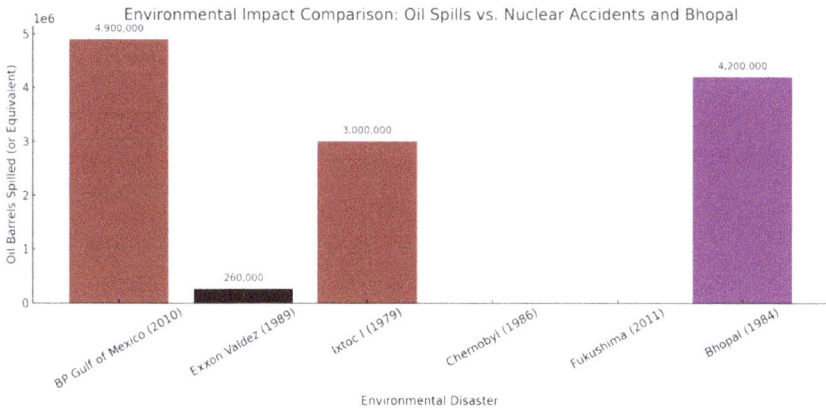

Environmental Impact Comparison: Oil Spills vs. Nuclear Accidents and Bhopal

Source: Own elaboration based on the data presented in the Summary Table at the end of this chapter

Graph comparing the impact of accidents in terms of barrels of oil spilled, highlighting the massive scale of oil disasters compared to nuclear accidents.

The Myth of Nuclear Waste Being Dangerous Forever

Nuclear waste is less impactful than waste from other industries.

It is not dumped into the environment, and its hazardous nature decreases over time.

While chemical waste remains toxic forever, nuclear waste is safely stored and can even be recycled.

The real question should not be, "How do we deal with nuclear waste?" but rather, "Why does no one worry about chemical and oil industry waste that truly destroys the environment?"

THE MYTH OF NUCLEAR WASTE BEING DANGEROUS FOREVER

MYTH

Nuclear waste is minimally impactful when compared to waste from other industries.

It is not dumped into the environment, and its hazardous nature decreases over time

While chemical waste remains toxic forever, nuclear waste is safely stored and can even be recycled

The real question should not be "how do we deal with nuclear waste?" but rather

why does no one worry about chemical and oil industry waste that truly destroys the environment?

Infographic debunking myths about nuclear waste, comparing misconceptions with real facts.

The Future of Waste Management – Alternatives and Innovations

The debate around nuclear energy often focuses on the issue of radioactive waste. However, recent technological advances are revolutionizing how this waste is processed and reused.

Contrary to common belief, nuclear waste does not have to be stored indefinitely. With new technologies, it is possible to reduce its hazardousness, repurpose it as fuel, and even eliminate it through nuclear transmutation.

In this section, we will explore the main innovations that could transform nuclear waste management in the 21st century.

The key to minimizing nuclear waste lies in reusing irradiated fuel. Rather than treating spent fuel as "waste," it can be reprocessed and transformed into a new energy source.

Two approaches are revolutionizing this concept:

Fast Reactors and Closing the Fuel Cycle

Fast Breeder Reactors (FBRs) are an advanced technology capable of utilizing spent nuclear fuel, making nuclear energy far more efficient and sustainable.

How do fast reactors work?

- Unlike conventional reactors, fast reactors do not require neutron moderators.

- They use fast neutrons to induce nuclear reactions, allowing nearly 100% of the fuel to be utilized.

- Uranium-238 and even plutonium-239 present in the waste can be reconverted into fissile fuel, drastically reducing long-lived waste.

Advantages of fast reactors:

- They can consume nuclear waste from other reactors, reducing the volume of radioactive waste.

- They increase uranium efficiency by up to 100 times, minimizing the need for mining.

- They generate less high-level radioactive waste.

Real-world example:

Russia is already operating the BN-600 and BN-800 fast reactors, which use recycled plutonium as fuel.

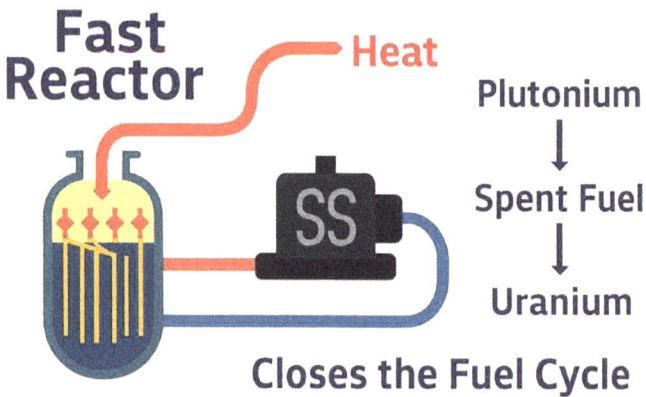

Diagram showing how a fast reactor works and how the fuel cycle is closed.

BN-600 and BN-800 Fast Reactors – The Future of Sustainable Nuclear Energy

Fast reactors are essential for the next generation of nuclear power plants, as they enable the reuse of irradiated fuel, reduce the need for uranium mining, and minimize long-lived waste.

In Russia, the BN (Bolshoy Moschnosty) series represents one of the most advanced and successful commercial systems in operation worldwide. Currently, the BN-600 and BN-800 reactors are the leading examples of this technology, operating at the Beloyarsk Nuclear Power Plant.

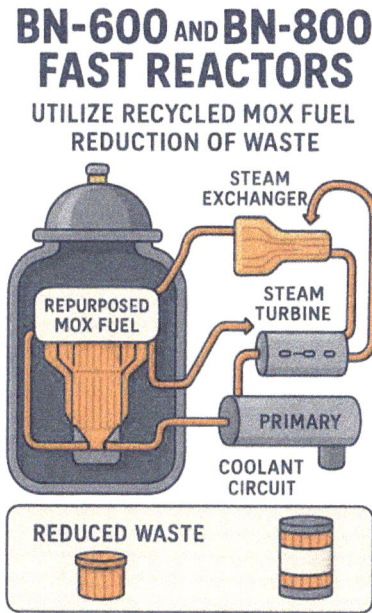

BN-600 AND BN-800 FAST REACTORS
UTILIZE RECYCLED MOX FUEL
REDUCTION OF WASTE

Diagram illustrating the operation of the BN-600 and BN-800 fast reactors, highlighting the use of recycled MOX fuel and the reduction of waste.

What Makes These Reactors Special?

- They use fast neutrons, allowing much greater efficiency in nuclear fission.

- They can consume plutonium and other nuclear waste, reducing the amount of long-lived radioactive waste.

- They are designed to close the nuclear fuel cycle, reusing uranium and plutonium from spent fuel.

- They use liquid sodium as a coolant, providing more efficient cooling without the need for neutron moderators.

1. The BN-600 Reactor – The Oldest and Most Reliable Commercial Fast Reactor

The BN-600 entered operation in 1980 at the Beloyarsk Nuclear Power Plant in Russia and is the oldest commercial fast reactor still operating in the world.

Technical Characteristics of the BN-600:

- Thermal power: 1,470 MWt

- Net electric power: 600 MWe

- Coolant: Liquid sodium

- Fuel type: MOX (Mixed Oxide Fuel) – a mixture of uranium and plutonium

What Makes the BN-600 So Important?

- It was the first commercially viable fast reactor, demonstrating that this technology can be safe and reliable.

- It has operated for over 40 years efficiently, accumulating a vast amount of operational data on fast reactors.

- It proved that nuclear fuel recycling and high-level waste reduction are achievable.

Interesting Fact: Despite being a technology developed during the Soviet Union era, the BN-600 continues to operate with a high-reliability rate.

2. The BN-800 Reactor – The New Generation of Fast Reactors

The BN-800 was built at the same Beloyarsk site and entered operation in 2015 as an enhanced version of the BN-600. It is considered the most advanced commercial fast reactor in the world today.

Technical Characteristics of the BN-800:

- Thermal power: 2,100 MWt

- Net electric power: 880 MWe

- Coolant: Liquid sodium

- Fuel type: MOX – Recycled plutonium + depleted uranium

What Makes the BN-800 So Innovative?

- It was designed to test the complete closure of the nuclear fuel cycle, using only recycled fuel.

- It demonstrates that fast reactors can operate with 100% MOX fuel, eliminating the need for additional uranium mining.

- It drastically reduces the production of high-level radioactive waste.

Interesting Fact: Since 2022, the BN-800 has been operating exclusively with recycled MOX fuel, becoming the first reactor in the world to achieve this milestone at a commercial scale.

RECYCLED MOX FUEL IN BN-600 AND BN-800
FAST REACTORS

PLUTONIUM

DEPLETED URANIUM

BENEFITS
- REUSES NUCLEAR WASTE
- REDUCES URANIUM MINING
- GENERATES LESS WASTE

USED IN BN-600 AND BN-800 FAST REACTORS

Infographic explaining how recycled MOX fuel is used in the BN-600 and BN-800 fast reactors, highlighting its composition and benefits.

3. Benefits of the BN-600 and BN-800 Fast Reactors:

- They allow nuclear waste to be reused as fuel, significantly reducing the amount of nuclear waste.

- They decrease the need for uranium mining, as they can operate with recycled fuel.

- They increase nuclear fuel efficiency by up to 100 times compared to conventional reactors.

- They reduce the risks of nuclear proliferation by transforming residual plutonium into energy instead of allowing its accumulation.

Conclusion: Russia is demonstrating that a closed nuclear fuel cycle is not just theoretical but a commercially viable reality.

Chart 36: Comparison of Efficiency Between Fast Reactors and Conventional Reactors

Efficiency Comparison: Fast Reactors vs. Conventional

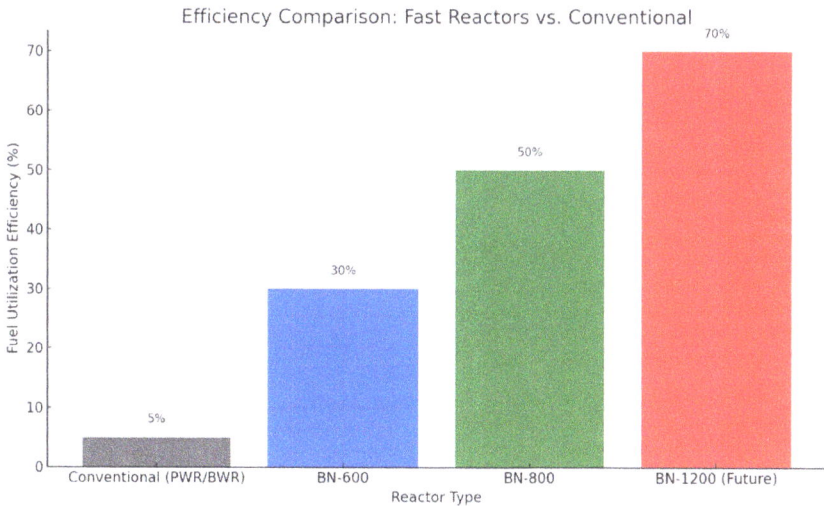

Source: Own elaboration based on the data presented in the Summary Table at the end of this chapter

Graph comparing the efficiency of fast reactors with conventional reactors, showing the significant increase in nuclear fuel utilization.

4. The Future – The BN-1200 and the Expansion of Fast Reactors

Following the success of the BN-600 and BN-800, Russia is already developing a new large-scale fast reactor, the BN-1200.

- Electric capacity of 1,200 MWe, making it one of the most powerful fast reactors in history.

- Improved safety and efficiency with recycled MOX fuel.

- Designed for export, potentially to be adopted by other countries.

If successfully implemented, the BN-1200 could consolidate nuclear energy as a completely sustainable system, eliminating the need for permanent geological repositories for high-level radioactive waste.

Conclusion – BN-600 and BN-800 Are the Path to Sustainable Nuclear Energy

The BN series fast reactors are proof that closing the nuclear fuel cycle is already a reality.

- The BN-600 demonstrated that fast reactors can operate safely and efficiently for over 40 years.

- The BN-800 achieved a historic milestone by operating 100% with recycled MOX fuel.

- Both reactors show that nuclear waste can be transformed into energy, reducing environmental impact.

- With the BN-1200, Russia plans to expand this technology on a global scale.

Thus, the idea that "nuclear waste is an unsolvable problem" is being refuted by the very technology that enables this material to be reused as fuel.

Final Reflection:

If all countries adopted fast reactors like the BN-600 and BN-800, long-lived nuclear waste would no longer be a concern, as it would be transformed into energy.

BN FAST REACTOR DEVELOPMENT

IMPROVE EFFICIENCY	USE MOX FUEL	ENHANCE SUSTAINABILITY

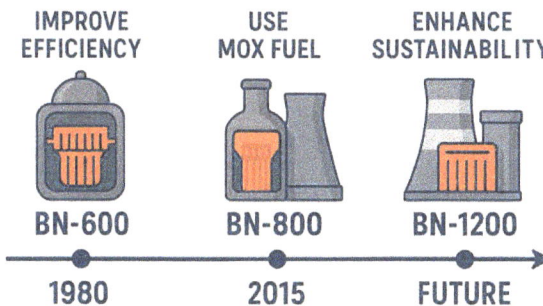

BN-600	BN-800	BN-1200
1980	2015	FUTURE

Timeline showing the evolution of the BN-600, BN-800, and BN-1200 fast reactors, highlighting advances in efficiency, use of MOX fuel, and sustainability.

Thorium Reactors and MSRs (Molten Salt Reactors)

Thorium reactors and Molten Salt Reactors (MSRs) represent an innovative approach that can transform nuclear waste into a source of energy.

What is Thorium?

- Thorium-232 is an abundant element in the Earth's crust and can be converted into Uranium-233, an excellent nuclear fuel.

- Unlike uranium, thorium produces less long-lived waste.

What are Molten Salt Reactors (MSRs)?

- They use liquid fuel dissolved in molten salts, allowing for greater safety and efficiency.

- Some MSRs can use plutonium and other nuclear waste as fuel, further reducing radioactive waste.

Advantages of Thorium Reactors and MSRs:

- Less nuclear waste and shorter radioactive decay times.

- Ability to consume existing radioactive waste, reducing the overall volume of nuclear waste.

- Passive safety system: no risk of core meltdown.

Real-world example:

China is developing an experimental thorium reactor, which could become the world's first commercially viable thorium reactor.

Thorium and Molten Salt Reactors (MSRs) – The Future of China's Nuclear Energy

Thorium and molten salt reactors are an advanced and safer alternative to traditional uranium-based reactors.

Thorium-232, the primary fuel for these systems, is far more abundant in the Earth's crust than uranium and produces much less long-lived radioactive waste.

China is at the forefront of developing these technologies, with its Experimental Molten Salt Reactor (TMSR-LF1) entering the testing phase in 2021, marking a major advancement for the nuclear sector.

Why Can Thorium Replace Uranium?

Thorium-232 is a fertile material that can be converted into Uranium-233, a highly efficient nuclear fuel.

Unlike Uranium-235, thorium:

- Is three to four times more abundant in the Earth's crust.

- Produces much less long-lived radioactive waste.

- Has a safer fuel cycle, reducing the risk of nuclear proliferation.

- Is highly efficient and can be almost completely utilized during fission.

Interesting fact: While 99% of natural uranium cannot be directly used as nuclear fuel, 100% of thorium can be converted into usable fuel.

How Do Molten Salt Reactors (MSRs) Work?

Molten Salt Reactors (MSRs) are an advanced type of nuclear reactor that uses liquid fuel dissolved in molten salts instead of solid uranium rods.

- Unlike conventional reactors, MSRs operate at much lower pressures, eliminating the risk of catastrophic explosions.

- If overheating occurs, the liquid fuel drains into a safety tank, automatically stopping the nuclear reaction.

- This technology allows the use of thorium, reducing dependence on uranium and minimizing the production of long-lived waste.

Key Benefits of MSRs:

- **Lower risk of accidents** – Liquid fuel cannot melt down like solid-fuel reactors.

- **Higher operating temperatures** – Greater efficiency in converting heat into electricity.

- **Low production of long-lived radioactive waste** – Much shorter decay times for radioactive materials.

- **Capability to consume plutonium and other nuclear waste**, reducing the volume of radioactive waste.

MOLTEN SALT REACTOR

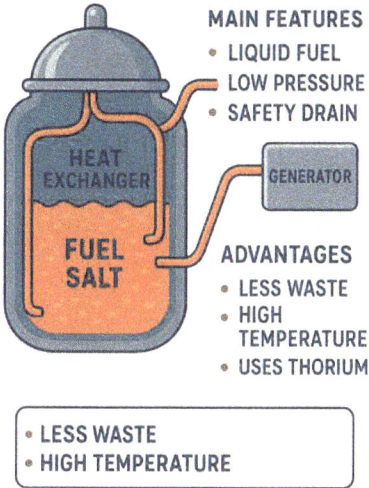

MAIN FEATURES
- LIQUID FUEL
- LOW PRESSURE
- SAFETY DRAIN

HEAT EXCHANGER

GENERATOR

FUEL SALT

ADVANTAGES
- LESS WASTE
- HIGH TEMPERATURE
- USES THORIUM

- LESS WASTE
- HIGH TEMPERATURE

Diagram illustrating how a molten salt reactor (MSR) works, highlighting its main features and advantages.

THORIUM AND MOLTEN SALT REACTORS (MSR)

THOREO

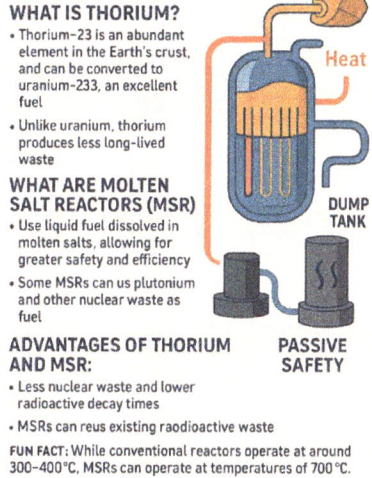

WHAT IS THORIUM?
- Thorium-23 is an abundant element in the Earth's crust, and can be converted to uranium-233, an excellent fuel
- Unlike uranium, thorium produces less long-lived waste

Heat

WHAT ARE MOLTEN SALT REACTORS (MSR)
- Use liquid fuel dissolved in molten salts, allowing for greater safety and efficiency
- Some MSRs can us plutonium and other nuclear waste as fuel

DUMP TANK

ADVANTAGES OF THORIUM AND MSR:
- Less nuclear waste and lower radioactive decay times
- MSRs can reus existing raodioactive waste

PASSIVE SAFETY

FUN FACT: While conventional reactors operate at around 300–400°C, MSRs can operate at temperatures of 700°C.

Infographic explaining thorium and molten salt reactors (MSR).

Interesting fact: While conventional reactors operate at around 300–400°C, MSRs can operate at temperatures above 700°C, making them much more efficient.

China's Experimental Molten Salt Reactor (TMSR-LF1)

China began operating an experimental molten salt reactor in Gansu Province in 2021, becoming the first country to test this technology at a real scale since the 1960s.

Characteristics of the TMSR-LF1 Reactor:

- Location: Gansu Province, Wuwei Desert
- Power: 2 MW thermal (prototype, with plans for 373 MW in the next phase)
- Fuel: Thorium dissolved in fluoride salts
- Objective: To validate the feasibility of MSRs for large-scale deployment

China's Next Steps:

- They plan to build a 373 MW reactor by 2030, enough to power a small city.
- Studies suggest the technology could be scaled up to 1 GW, making it a real alternative to uranium-based power plants.
- China has invested billions of dollars into its thorium program, betting that this will be the foundation of future nuclear energy.

Infographic explaining the advantages of thorium as a nuclear fuel, highlighting its abundance, safety, and lower waste generation.

If successful, this technology could replace conventional reactors, making nuclear energy much safer, cheaper, and more sustainable.

Comparison Between Molten Salt Reactors and Conventional Reactors

Molten salt reactors offer significant advantages over conventional pressurized water reactors (PWR/BWR).

Table 34: Comparison Between Conventional Reactors and New Molten Salt Reactors

Feature	Conventional Reactors (PWR/BWR)	Molten Salt Reactors (MSR)
Fuel	Enriched uranium (solid)	Thorium dissolved in molten salts
Operating pressure	High pressure (150 atm)	Low pressure (near ambient)
Core meltdown risk	High, if the cooling system fails	Extremely low since the fuel is already molten
Thermal efficiency	33-35%	45-50%
Long-lived waste	Produces plutonium and actinides	Much less long-lived radioactive waste
Nuclear proliferation	Possible, as it can generate weapons-grade plutonium	Virtually impossible, as U-233 is hard to divert

Source: Own elaboration based on the data presented in the Summary Table at the end of this chapter

Conclusion: MSRs could revolutionize nuclear energy by offering a much safer, more efficient system with a lower environmental impact.

Chart 37: Comparison of Efficiency Between Molten Salt Reactors and Conventional Reactors

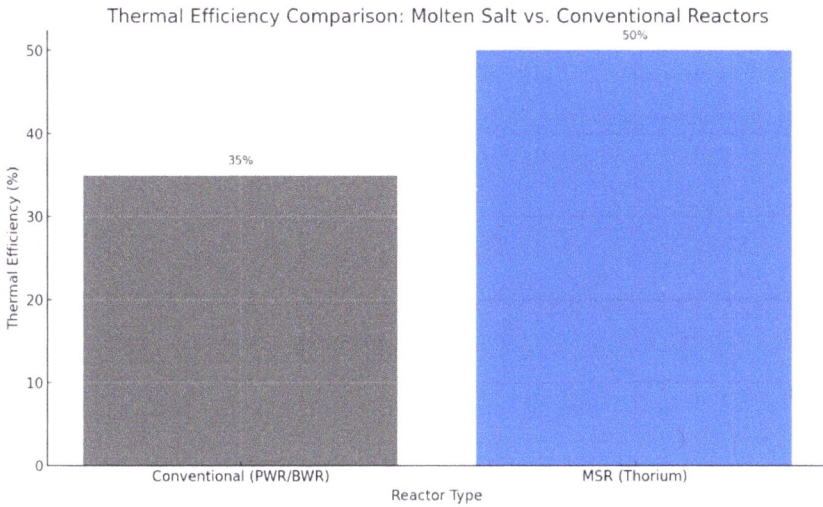

Thermal Efficiency Comparison: Molten Salt vs. Conventional Reactors

Source: Own elaboration based on the data presented in the Summary Table at the end of this chapter

Graph comparing the efficiency of MSRs with conventional reactors, highlighting the superiority of molten salt reactors in terms of energy utilization.

The Future of Thorium and MSR Reactors:

- China is leading the race for molten salt reactors, with plans for commercial reactors by 2030.

- Other countries such as the USA, Canada, and India are also investing in the technology.

- In the long term, MSRs could completely replace traditional uranium reactors, making nuclear energy safer and more sustainable.

Final Reflection: If this technology succeeds, nuclear energy could become virtually inexhaustible and free from the problems associated with long-lived nuclear waste.

Conclusion – Thorium and MSRs Could Revolutionize Nuclear Energy:

- Thorium is abundant and can replace uranium in power generation.

- Molten salt reactors are much safer and more efficient than conventional reactors.

- China is leading the development of this technology and could transform the global nuclear market.

- If widely adopted, MSRs could eliminate the problem of long-lived waste, making nuclear energy even more sustainable.

EVOLUTION OF CHINA'S THORIUM AND MSR PROJECT

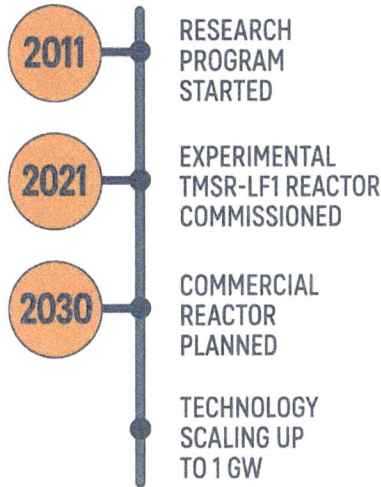

2011 — RESEARCH PROGRAM STARTED

2021 — EXPERIMENTAL TMSR-LF1 REACTOR COMMISSIONED

2030 — COMMERCIAL REACTOR PLANNED

TECHNOLOGY SCALING UP TO 1 GW

Timeline showing the evolution of China's thorium and MSR project, highlighting key milestones in the development of this technology.

Thus, thorium and MSR reactors represent one of the most promising paths for the future of clean and safe energy.

Nuclear Transmutation Technologies to Reduce Waste Longevity

One of the most promising technologies for waste management is nuclear transmutation, which can convert highly radioactive elements into short-lived or even non-radioactive materials.

How does nuclear transmutation work?

- Radioactive waste is exposed to fast neutrons in reactors or particle accelerators.

- This process alters the isotopes of the elements, significantly reducing the time needed for the waste to become harmless.

Benefits of nuclear transmutation:

- Reduces the longevity of waste from thousands of years to just centuries or decades.

- Can be integrated into fast reactors and particle accelerators.

- Decreases the need for permanent geological storage.

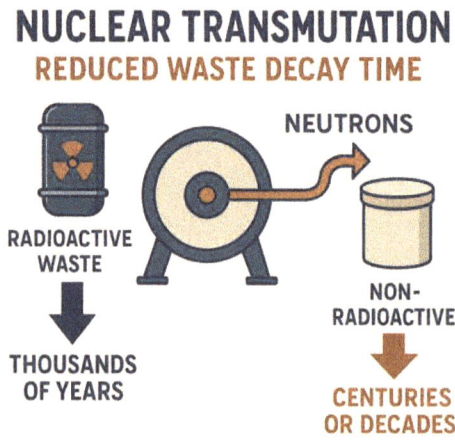

NUCLEAR TRANSMUTATION
REDUCED WASTE DECAY TIME

NEUTRONS

RADIOACTIVE WASTE

THOUSANDS OF YEARS

NON-RADIOACTIVE

CENTURIES OR DECADES

Diagram showing how nuclear transmutation can reduce the decay time of radioactive waste.

Real-world example: The European MYRRHA project in Belgium is testing nuclear transmutation to eliminate high-level radioactive waste.

MYRRHA – The Future of Nuclear Transmutation and Waste Reduction

MYRRHA (Multi-purpose Hybrid Research Reactor for High-tech Applications) is an experimental hybrid reactor developed by Belgium and supported by the European Union.

It combines a fast nuclear reactor using liquid metals with a particle accelerator, enabling research in:

- Nuclear transmutation to reduce the decay time of nuclear waste.

- Production of medical isotopes for cancer diagnosis and treatment.

- Testing of new nuclear fuels for future reactors.

This project is unique in the world and could revolutionize how nuclear waste is managed.

1. How Does MYRRHA Work?

MYRRHA is a hybrid reactor, meaning it is not a conventional nuclear reactor. Instead, it is a subcritical reactor that depends on a particle accelerator to operate.

- **Particle accelerator**: MYRRHA uses a high-energy proton beam to maintain the nuclear reaction.

- **Subcritical reactor**: Without the proton beam, the reactor simply shuts down, making it extremely safe.

- **Use of lead-bismuth coolant**: Unlike conventional reactors, which use pressurized water, MYRRHA uses a mixture of molten lead and bismuth, increasing both safety and efficiency.

- **Nuclear transmutation**: MYRRHA can transform highly radioactive waste into short-lived elements, drastically reducing the time needed for waste to become safe.

This technology could eliminate the need to store nuclear waste for hundreds of thousands of years, cutting the timescale down to just a few decades or centuries.

HOW MYRRHA WORKS

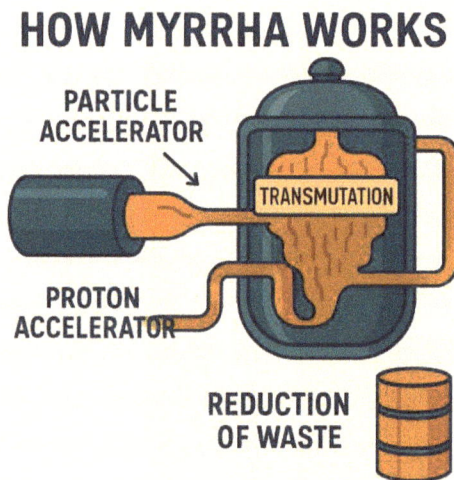

Diagram illustrating how MYRRHA works and its role in nuclear transmutation, showing its particle accelerator, cooling system, and benefits in waste reduction.

2. Objectives of the MYRRHA Project

MYRRHA was designed to test new solutions for the future of nuclear energy. Its main objectives are:

Reduce the Longevity of Nuclear Waste

- Nuclear transmutation can reduce the decay time of highly radioactive materials from 100,000 years to less than 300 years.

- This eliminates the need for permanent geological repositories for much of the nuclear waste.

Develop New Technologies for Future Reactors

- MYRRHA tests new types of nuclear fuels and advanced coolants (such as lead-bismuth).

- The data collected will support the development of next-generation fast reactors.

Produce Medical Isotopes for Diagnostics and Treatments

- MYRRHA can produce essential medical isotopes, such as Molybdenum-99, used in cancer treatments.

- This could reduce Europe's dependence on aging nuclear reactors for the production of these materials.

MYRRHA not only addresses the problem of nuclear waste but also advances medicine and sustainable nuclear technology.

DEVELOPMENT OF THE MYRRHA PROJECT

Timeline showing historical milestones
and advances in nuclear transmutation

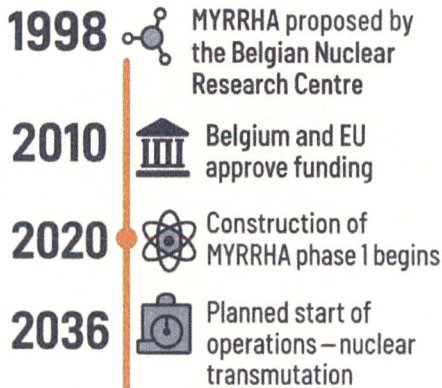

1998 — MYRRHA proposed by the Belgian Nuclear Research Centre

2010 — Belgium and EU approve funding

2020 — Construction of MYRRHA phase 1 begins

2036 — Planned start of operations — nuclear transmutation

Timeline showing the development of the MYRRHA project, highlighting its historical milestones and advances in nuclear transmutation.

3. How Nuclear Transmutation Works in MYRRHA:

Nuclear transmutation is a process in which highly radioactive elements are bombarded with fast neutrons, altering their atomic structure and transforming them into short-lived or non-radioactive elements.

In MYRRHA, this occurs as follows:

1. The particle accelerator generates a high-energy proton beam.

2. This beam strikes a lead-bismuth target, producing fast neutrons.

260

3. The neutrons bombard nuclear waste, breaking their nuclei and transforming them into short-lived isotopes.

The result?

- Elements that would take 100,000 years to decay become safe in just a few hundred years.

- Drastic reduction in the need for permanent geological storage.

Important fact: Nuclear transmutation can enable 90% of high-level radioactive waste to be converted into harmless elements.

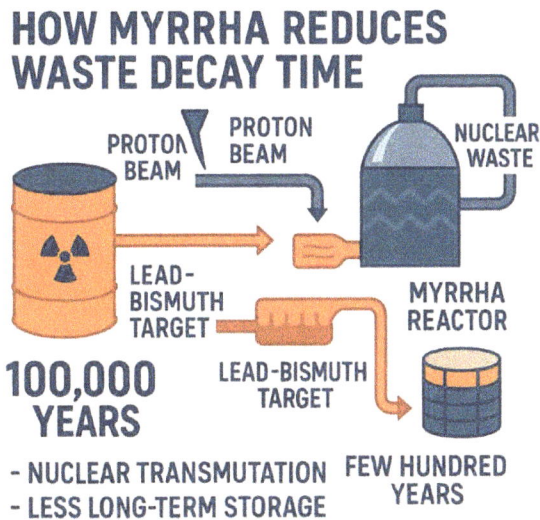

Infographic explaining how MYRRHA reduces the decay time of radioactive waste through nuclear transmutation, highlighting the benefits for waste management and environmental impact.

4. The Future of MYRRHA and Its Impact on Nuclear Energy

MYRRHA is being developed in three phases, with completion expected by 2040.

Phase 1 (2027): Construction of the 100 MeV particle accelerator for initial testing.

Phase 2 (2033): Expansion to 600 MeV, enabling nuclear transmutation experiments.

Phase 3 (2040): Full construction of the subcritical hybrid reactor, capable of large-scale operation.

If successful, MYRRHA could lead to the creation of new nuclear power plants that not only produce energy but also eliminate radioactive waste.

Conclusion: This project could completely transform the perception of nuclear energy, making it even safer and more sustainable.

5. Comparison Between MYRRHA and Conventional Reactors

Table 35: Comparison Between MYRRHA and Conventional Reactors

Feature	Conventional Reactors (PWR/BWR)	MYRRHA (Hybrid Transmutation Reactor)
Fuel	Enriched uranium	Can use nuclear waste as fuel
Operating pressure	High pressure (150 atm)	Low pressure (lead-bismuth coolant)

Waste management	Produces large volumes of long-lived waste	Reduces or eliminates high-radioactivity waste
Safety	Core meltdown risk	Extremely safe – shuts down automatically without particle accelerator
Medical applications	None	Produces essential medical isotopes

Source: Own elaboration based on the data presented in the Summary Table at the end of this chapter

MYRRHA not only produces energy but also addresses the nuclear waste problem, making it one of the most innovative projects today.

Conclusion – MYRRHA Is the Key to Sustainable Nuclear Energy

- Nuclear transmutation can drastically reduce the decay time of radioactive waste.

- MYRRHA is an extremely safe subcritical hybrid reactor, as it can be shut down instantly.

- The project will enable new advances in nuclear medicine and energy generation.

- If successful, it could eliminate the need for permanent geological repositories for nuclear waste.

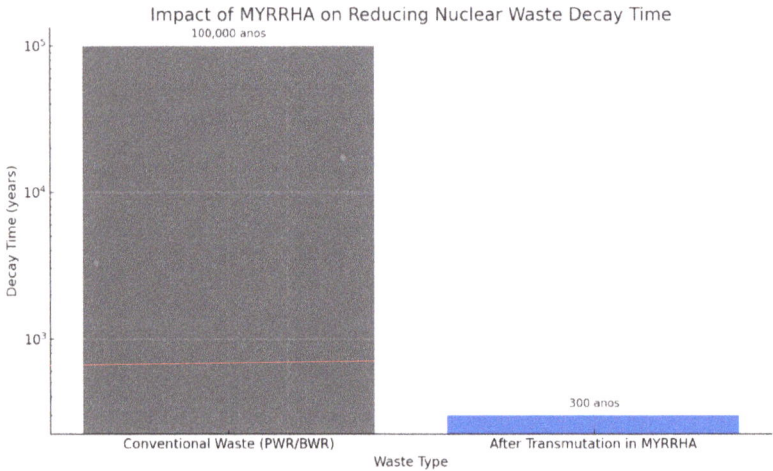

Impact of MYRRHA on Reducing Nuclear Waste Decay Time

Source: Own elaboration based on the data presented in the Summary Table at the end of this chapter

Graph showing the impact of MYRRHA on radioactive waste management, highlighting the significant reduction in waste decay time after nuclear transmutation.

Thus, MYRRHA could be a game-changer for the future of nuclear energy in Europe, demonstrating that nuclear waste is not an unsolvable problem but rather an opportunity for innovation.

For a region that spent decades paralyzed by the activism of "No to Nuclear" movements—falling significantly behind major Asian competitors like Russia and China—there is now an opportunity to make up for lost time with this innovative system.

It is unfortunate, however, that the process is taking so long.

Major Nuclear Innovation Programs in the United States

1. ARDP (Advanced Reactor Demonstration Program):

The Advanced Reactor Demonstration Program (ARDP) is one of the U.S. Department of Energy's main initiatives to accelerate the development of advanced and sustainable nuclear reactors in the United States.

Launched in 2020, the program has invested over $3 billion in new technologies.

Objective: To build advanced nuclear reactors for commercialization by the 2030s.

Two companies were selected to lead the first phase of the project:

- X-Energy – High-Temperature Gas-Cooled Reactor (HTGR)

- TerraPower – Natrium Reactor (Liquid Sodium-Cooled)

The U.S. is betting on high-temperature and sodium-cooled reactors, which could be safer and more efficient than current models.

2. TerraPower – The Natrium Reactor

TerraPower, founded by Bill Gates, is developing the Natrium reactor, a fast reactor cooled by liquid sodium that promises to be safer and more efficient than conventional reactors.

Advantages of the Natrium Reactor:

- Uses liquid sodium as a coolant, reducing the risk of core meltdown.

- Can operate with recycled fuel, reducing the need for uranium mining.

- Features a thermal energy storage system, allowing flexibility in electricity generation.

Interesting fact: The first plant with the Natrium reactor is being built in the state of Wyoming and is expected to be operational by 2030.

HOW THE TERRAPOWER **NATRIUM REACTOR WORKS**

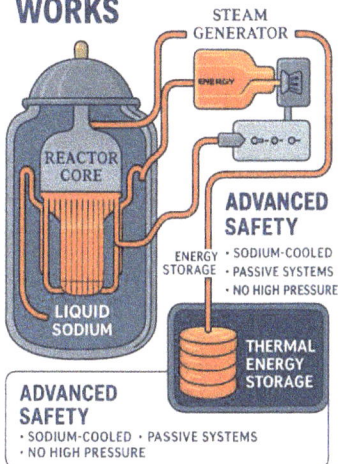

STEAM GENERATOR

ENERGY

REACTOR CORE

ADVANCED SAFETY

ENERGY STORAGE
· SODIUM-COOLED
· PASSIVE SYSTEMS
· NO HIGH PRESSURE

LIQUID SODIUM

THERMAL ENERGY STORAGE

ADVANCED SAFETY
· SODIUM-COOLED · PASSIVE SYSTEMS
· NO HIGH PRESSURE

Diagram illustrating how TerraPower's Natrium reactor works, highlighting its sodium cooling system, energy storage, and advanced safety features.

X-ENERGY
SMALL MODULAR REACTORS

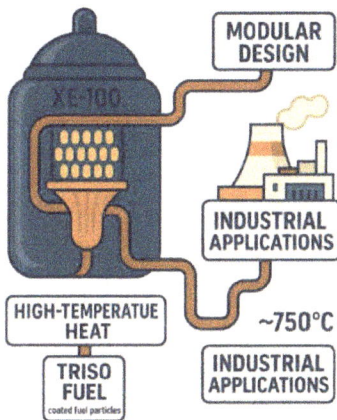

XE-100

MODULAR DESIGN

INDUSTRIAL APPLICATIONS

HIGH-TEMPERATUE HEAT

~750°C

TRISO FUEL
coated fuel particles

INDUSTRIAL APPLICATIONS

Infographic showing X-Energy's Small Modular Reactors (SMRs), highlighting their modular design, TRISO fuel, and industrial applications.

3. X-Energy – High-Temperature Small Modular Reactors (SMRs)

X-Energy is developing a high-temperature small modular reactor (SMR) called the Xe-100, which can be used to provide direct heat for heavy industries in addition to generating electricity.

Characteristics of the Xe-100:

- Operates at extremely high temperatures (~750°C), allowing for greater thermal efficiency.

- Uses TRISO fuel, one of the safest fuels in the world, resistant to meltdown.

- Modular design – Small, safe, and cost-effective.

Conclusion: X-Energy's SMRs could be a solution for decarbonizing heavy industries, such as steel production and hydrogen generation.

4. Oklo – The Compact and Self-Sustaining Fission Reactor

The startup Oklo is developing a compact reactor called Aurora, designed to operate for decades without needing refueling.

Key Features:

- Uses recycled depleted uranium, reducing nuclear waste.

- Passive design – No pumps or active systems are needed to prevent accidents.

- Capable of providing electricity to remote locations and military bases.

Interesting fact: Aurora is designed to operate for 20 years without refueling, making it a promising solution for providing stable energy in isolated areas.

5. Nuclear Fusion Projects – Opportunity for Infinite Clean Energy

The United States is also heavily investing in nuclear fusion, a technology that could revolutionize energy generation.

Major fusion projects in the U.S.:

- National Ignition Facility (NIF) – The first laboratory to achieve fusion ignition in 2022.

- Commonwealth Fusion Systems (CFS) – A startup developing advanced tokamaks with superconducting magnets.

- Helion Energy – Developing an innovative pulsed fusion system to generate electricity.

If nuclear fusion is successfully mastered, the U.S. could become a leader in clean, limitless energy without nuclear waste.

HOW NUCLEAR FUSION WORKS

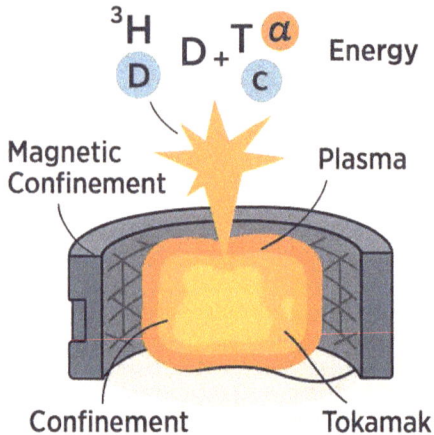

^{3}H

D

D + T α
C

Energy

Magnetic
Confinement

Plasma

Confinement

Tokamak

**Advantages Over
Nuclear Fission**

Fuel abundant,
easy to obtain:
seawater

No risk of accident:
no chain reaction

Little waste:
product is helium

Fuel abundant,
easy to obtain

*Diagram illustrating how nuclear fusion works and its benefits, highlighting
the deuterium-tritium reaction, magnetic confinement, and advantages
over nuclear fission.*

6. Fourth-Generation Reactors – The Future of Nuclear Energy in the U.S.

In addition to current technologies, the U.S. is conducting research into fourth-generation reactors, which include:

- High-temperature gas-cooled Reactors (HTGRs) – Safe and efficient, capable of producing hydrogen as a byproduct.

- Molten Salt Reactors (MSRs) – No risk of core meltdown and capable of operating with thorium.

- Fluoride-salt-cooled high-temperature Reactors (FHRs) – A combination of graphite reactors and molten salt technology.

With these advances, the U.S. could completely transform the nuclear industry by 2050, making it safer, more sustainable, and more efficient.

The U.S. Is Leading the Next Nuclear Revolution

- The U.S. is developing next-generation reactors that are safer and more efficient.

- The government is funding projects to accelerate the nuclear transition.

- Technologies like nuclear fusion and molten salt reactors could revolutionize the sector.

- By 2050, the U.S. could have an entirely innovative and sustainable nuclear system.

Timeline of U.S. Nuclear Programs

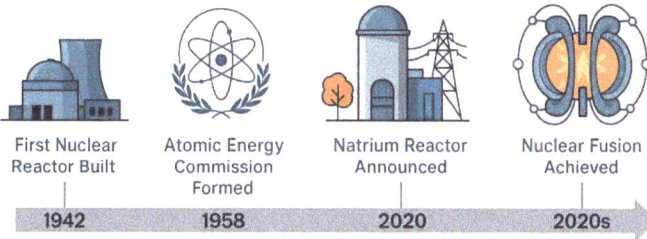

First Nuclear Reactor Built	Atomic Energy Commission Formed	Natrium Reactor Announced	Nuclear Fusion Achieved
1942	1958	2020	2020s

Timeline showing the evolution of U.S. nuclear programs, highlighting major milestones from the first reactor to recent innovations like Natrium and nuclear fusion

Timeline showing the evolution of U.S. nuclear programs, highlighting major

milestones from the first reactor to the latest innovations such as Natrium and nuclear fusion.

With these initiatives, the United States can continue to be one of the most influential countries in the nuclear sector, ensuring a clean and reliable energy source for the future.

Unlike Europeans, the U.S. never embraced the narratives portraying nuclear energy as the 'devil' that should be completely eliminated.

Although progressing at a slower pace compared to its main competitors, Russia and China, the U.S. never lost focus on mastering and innovating in this critical field.

They always understood that the ability to produce abundant and low-cost energy places a country in a much stronger position to face the challenges of economic cycles and to provide more sustainable economic development.

With the newly inaugurated Trump Administration announcing its special focus on the energy sector — and on nuclear energy in particular — it is likely that, within a few years, the United States will once again lead this fundamental sector.

Future Outlook: Can Nuclear Energy Become a Sustainable System?

With all these innovations, it is possible to envision a future where nuclear waste is no longer a problem but rather a new energy resource.

Scenario for the Future of Nuclear Energy:

- Closing the fuel cycle – Fast reactors and nuclear transmutation will allow nearly 100% of uranium to be utilized, drastically reducing waste.

- Use of thorium and advanced reactors – Replacing uranium with thorium could make waste much less problematic while reducing nuclear proliferation risks.

- Recycling of nuclear fuel – Countries like France already reprocess over 80% of their used nuclear fuel. In the future, this figure could approach 100%.

- Drastic reduction of long-lived waste – Through nuclear transmutation, the most dangerous wastes could have their decay time reduced from hundreds of thousands of years to just a few decades.

If advanced reactors, fuel recycling, and nuclear transmutation are combined, nuclear energy could become virtually sustainable, eliminating the need for permanent geological repositories.

Table 36: Emerging Technologies in Waste Management

Technology	Operating Principle	Waste Reduction Potential
Fast Reactors	Use plutonium/depleted uranium as fuel	Reduce waste volume by up to 90%
Molten Salt Reactors	Liquid fuel enables high safety and efficiency	Can reuse legacy nuclear waste
Nuclear Transmutation	Converts long-lived isotopes into shorter-lived ones	Minimizes need for long-term storage

Conclusion of the Current Chapter – The Future of Nuclear Waste Management Is Promising

- Nuclear fuel can be reused almost indefinitely with fast reactors and recycling technologies.

- Nuclear waste can be converted into fuel for new power plants.

- Nuclear transmutation can drastically reduce the time required for waste to become harmless.

- Nuclear energy has the potential to become one of the most sustainable long-term energy sources.

Chart 39: Waste Reduction and Growth of Nuclear Energy

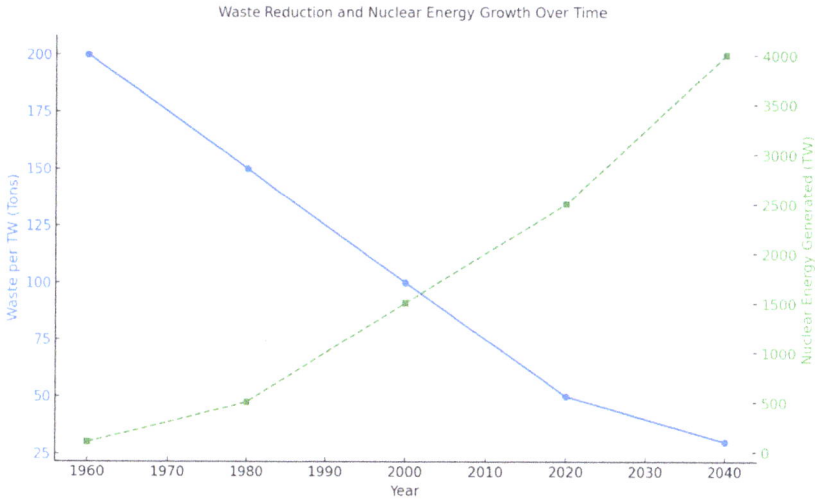

Waste Reduction and Nuclear Energy Growth Over Time

Source: Own elaboration based on the data presented in the Summary Table at the end of this chapter

This graph shows that despite the exponential growth of nuclear energy, the amount of waste generated per unit of energy has significantly decreased, thanks to technological advances and fuel recycling.

Contrary to the traditional narrative, nuclear waste does not have to be an eternal problem. With innovation, investment, and research, we can eliminate high-level radioactive waste and transform nuclear energy into a clean and sustainable system.

Table 37: Sources Consulted in Chapter 5

Source	Description
World Nuclear Association (WNA)	Data on radioactive waste volumes and management technologies.
International Atomic Energy Agency (IAEA)	Official reports on waste management, reactor technology, and nuclear transmutation.
OECD Nuclear Energy Agency (NEA)	Publications on nuclear fuel recycling and advanced reactor development.
European Commission (EC)	Research programs related to MYRRHA and nuclear innovation in Europe.
U.S. Department of Energy (DOE)	Information on ARDP, Natrium Reactor, X-Energy, and fusion projects.
TerraPower	Technical data and updates on the Natrium fast reactor project.
X-Energy	Technical documentation on the Xe-100 small modular reactor (SMR).
Oklo Inc.	Information about the Aurora reactor and compact fission systems.
Commonwealth Fusion Systems	Research on advanced magnetic confinement fusion (tokamaks).
Helion Energy	Developments in pulsed nuclear fusion systems for electricity generation.

International Thermonuclear Experimental Reactor (ITER)	Information on large-scale international nuclear fusion projects.
MIT Energy Initiative	Studies and projections for the future of nuclear energy and waste management.

Preparation for the Next Chapter – The False Anti-Nuclear Arguments and Their Motivations

Despite all the scientific evidence, technological advances, and the increasing safety of nuclear energy, the anti-nuclear movement continues to influence political decisions and public opinion.

But why?

If nuclear energy is safer than many industries, has a lower environmental impact than fossil fuels, and can provide clean and reliable electricity, what truly motivates the fierce opposition to this technology?

Throughout history, various myths and distorted arguments have been used to justify the fear of nuclear energy. Many of these criticisms are not based on facts but rather on political, economic, and ideological interests.

In the next chapter, we will analyze:

- The main anti-nuclear arguments and their inconsistencies.

- Who really funds the anti-nuclear movement?

- Why do some governments and companies prefer to boycott nuclear energy?

- How disinformation campaigns have shaped public perception of nuclear energy.

By dismantling these myths and exposing the true motivations behind the anti-nuclear movement, we will see that the resistance against nuclear energy is not merely a matter of safety or environment but rather one of politics, economics, and the manipulation of public opinion.

Now, let us examine what lies behind the opposition to nuclear energy and how these misconceptions are hindering the development of one of the most efficient and sustainable energy sources on the planet.

Chapter 6 -
The False Anti-Nuclear Arguments and Their Motivations

Nuclear energy is, without a doubt, one of the most misunderstood technologies of our time. Despite its proven safety, energy efficiency, and low environmental impact, it continues to be the target of fierce criticism and misinformation, perpetuated both by ignorance and by hidden political and economic interests.

The fear of nuclear energy did not arise spontaneously. On the contrary, it was cultivated over the decades through alarmist narratives that associate this form of energy with destruction, danger, and catastrophes. From the horrors of the Hiroshima and Nagasaki bombs to the accidents at Chernobyl and Fukushima, nuclear energy has been demonized disproportionately compared to its true impact.

Curiously, this rejection does not hold up when confronted with concrete data. While nuclear accidents are extremely rare and of limited impact, other industrial sectors, such as oil, coal, and chemical industries, cause much more devastating environmental and human tragedies – and almost never generate the same level of public outrage.

Who benefits from the fear of nuclear energy?

The answer to this question leads us to a complex network of political, ideological, and financial interests.

The Fossil Fuel Industry: Oil, gas, and coal have dominated the global energy sector for over a century. For these industries, the expansion of nuclear energy represents a direct threat, as it offers a reliable and low-carbon alternative. Keeping the public afraid of nuclear energy helps ensure that fossil fuels remain the foundation of the global energy supply.

Environmental NGOs and Political Movements: Paradoxically, many environmental groups that claim to fight climate change oppose nuclear energy, even knowing that it is one of the cleanest and most reliable sources of electricity. Many of these organizations receive funding from governments and companies interested in promoting intermittent renewable energies (such as solar and wind), which cannot completely replace nuclear generation.

Sensationalist Media: Catastrophes and alarmism sell newspapers and television hours, generate clicks, and dominate political debates. A single isolated nuclear accident, even without fatalities, can cause worldwide panic, while environmental disasters caused by oil and coal often go unnoticed.

Governments and Geopolitics: Nuclear energy means energy independence. Countries that develop nuclear plants reduce their dependence on importing gas and oil, something that does not always interest the major energy-exporting powers such as Russia, Saudi Arabia, and even the USA.

Thus, it is no surprise that anti-nuclear campaigns have been financed and supported by external interests throughout history.

How was the public manipulated?

If nuclear energy is safe, efficient, and necessary, why do so many people still believe it is a danger?

The answer lies in the systematic disinformation that has been spread since the 1960s.

- **The Fear Industry**: Nuclear fear was amplified by films, series, and sensationalist news, always portraying the technology as something unstable and apocalyptic.

- **'The China Syndrome' and the Hollywood Effect**: The film 'The China Syndrome' (1979) was released two weeks before the Three Mile Island accident and helped cement the idea that nuclear energy was an imminent disaster. Since then, Hollywood has used and abused this fear in productions like Chernobyl (HBO), The Simpsons, and countless post-apocalyptic movies.

- **Data Distortion**: Deaths attributed to nuclear energy are exaggerated or manipulated, while the impacts of other energy sectors are minimized or ignored.

- **Politics and Excessive Regulation**: Anti-nuclear pressure has led to an extreme increase in bureaucratic barriers to the construction of new plants, artificially raising the cost of nuclear energy and hindering its expansion.

The Purpose of This Chapter

In the next topics, we will dismantle, one by one, the main anti-nuclear arguments, showing what is true and what is pure manipulation.

We will answer questions such as:

- Is nuclear energy really dangerous?

- What happens to nuclear waste?

- Can renewable energies replace nuclear?

- Who is behind the anti-nuclear movement?

The reality is that the fear of nuclear energy is not based on science but rather on decades of disinformation. And it is time to change this narrative.

Myths and False Arguments Against Nuclear Energy

"Nuclear Energy is Dangerous" – Comparison with Other Industries:

The claim that "nuclear energy is dangerous" is one of the most persistent and widely accepted myths by the public, but also one of the easiest to refute with concrete data.

The truth is that nuclear energy is among the safest forms of electricity generation in the world. Its risks are extremely low compared to other industries that operate without the same level of oversight and control.

Let us dismantle this myth through a direct comparison with other forms of energy generation and industrial activities.

What Does "Danger" Mean in Energy Production?

When we talk about "danger" in energy generation, we can analyze the following factors:

1. Number of direct and indirect deaths caused by the industry over time.

2. Environmental impact and long-term effects on human health.

3. Risk of accidents and magnitude of the consequences.

Nuclear energy is often associated with catastrophic accidents, but if we look at the numbers, we will see that it kills fewer people than any other energy source.

Comparison of Mortality by Energy Source:

A study by Our World in Data (2022) analyzed the number of deaths per Terawatt-hour (TWh) of electricity generated, considering direct and indirect impacts, such as air pollution and industrial accidents.

Here are the results:

Chart 40: Comparison of Fatalities Among Different Energy Sources

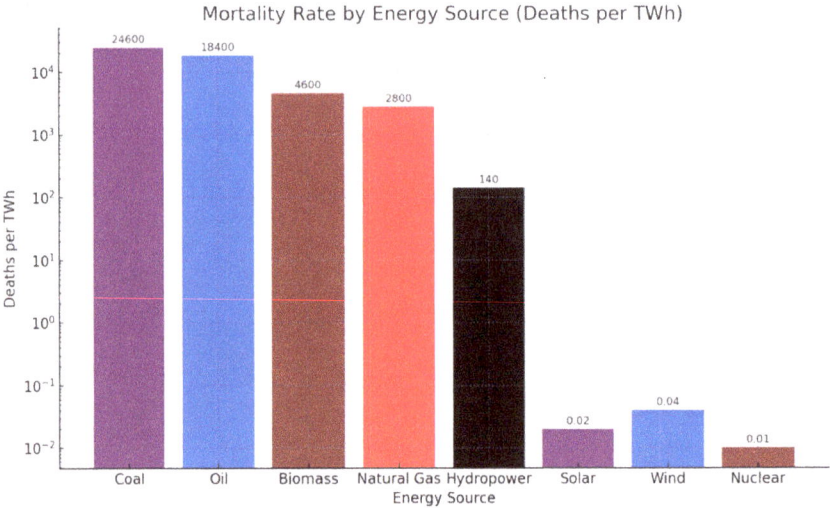

Source: Own elaboration based on the data presented in the Summary
Table at the end of this chapter

A comparative graph of mortality per TWh across different energy sources
clearly shows that nuclear energy is one of the safest in the world.

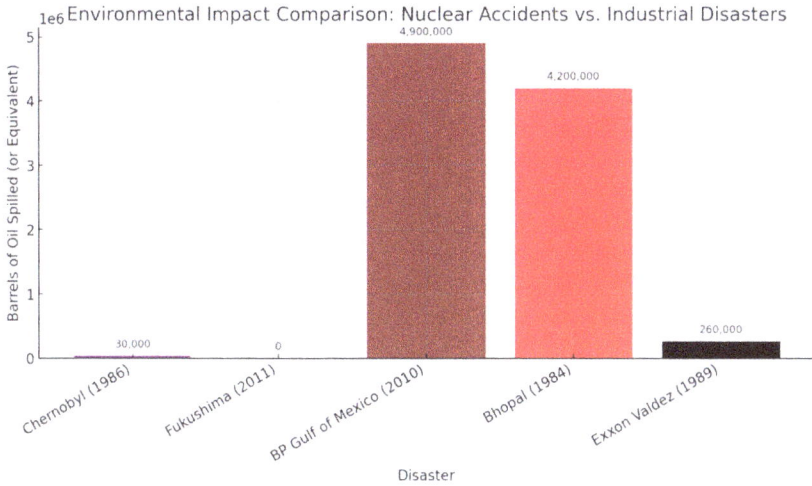

Source: Own elaboration based on the data presented in the Summary Table at the end of this chapter

Graph comparing the environmental impact of nuclear accidents vs. industrial and fossil fuel disasters. It shows that, despite the hype around Chernobyl and Fukushima, chemical and oil disasters had a much greater environmental impact.

Conclusion:

- Nuclear energy is the safest energy source in the world, surpassing Wind and Solar.

- It is 2,800 times safer than coal, which still represents a significant share of the global energy matrix.

- The risks of nuclear energy are statistically irrelevant when compared to the damage caused by fossil fuels.

But What About Nuclear Accidents? What About Chernobyl and Fukushima?

Opponents of nuclear energy often use the accidents at Chernobyl and Fukushima as arguments to justify that nuclear energy is dangerous. However, this narrative ignores the context, the evolution of nuclear safety, and the true consequences of these events.

- **Chernobyl (1986)** – An unsafe reactor design and human failures led to the worst nuclear accident in history. The RBMK (Graphite-Moderated Reactor) had no containment, and operators ignored safety procedures.

- **Fukushima (2011)** – A tsunami of historic proportions caused the disaster, and even so, no one died from direct radiation exposure. The impact was much smaller than that of fossil fuel disasters, such as the BP oil spill in the Gulf of Mexico (2010).

"Nuclear Waste is an Insoluble Problem"

This is undoubtedly one of the arguments most frequently used by critics of nuclear energy:

"Nuclear energy cannot be considered clean because we don't know what to do with the waste."

But is this really true? Or are we facing yet another persistent myth fueled by misinformation and ignorance about the technological advances already available?

The Truth: Nuclear Waste Management is Technically and Scientifically Solved

Contrary to popular belief, nuclear waste is highly controlled, rigorously managed, and occupies minimal volumes compared to the waste generated by other industries.

Worldwide, radioactive waste is classified into:

- Low-level waste (clothing, tools, filters – 90% of total volume);
- Intermediate-level waste (resins, reactor components);
- High-level waste (used fuel).

Most of the waste (low and intermediate-level) loses its radioactivity within decades or a few centuries and can be safely stored on the surface or in intermediate repositories.

High-level waste, which represents less than 3% of the total volume, is today safely stored, and completely viable long-term solutions exist, such as deep geological storage (example: Onkalo, in Finland).

The Myth of "Eternal Danger"

One of the most repeated arguments is that nuclear waste "remains dangerous for hundreds of thousands of years." What is often omitted, however, is that:

- Most of the radioactivity from the waste decays rapidly within the first few decades.

- After about 300 to 500 years, the radiation level of the waste becomes comparable to that of natural uranium ores in the earth's crust.

- Nuclear transmutation technologies already allow for drastically reducing the lifetime of the most hazardous waste (as we saw in Chapter 5 with the MYRRHA project).

Conclusion: The waste issue is not technical; it is political and psychological.

Let us observe the following graph, which shows the volume of toxic or hazardous waste generated per energy source to produce 1 TWh of electricity — including nuclear, coal, solar, and natural gas — to demonstrate that nuclear energy is, paradoxically, one of the technologies that generate the least waste per energy produced.

Chart 42: Volume of Waste Generated by Energy Source

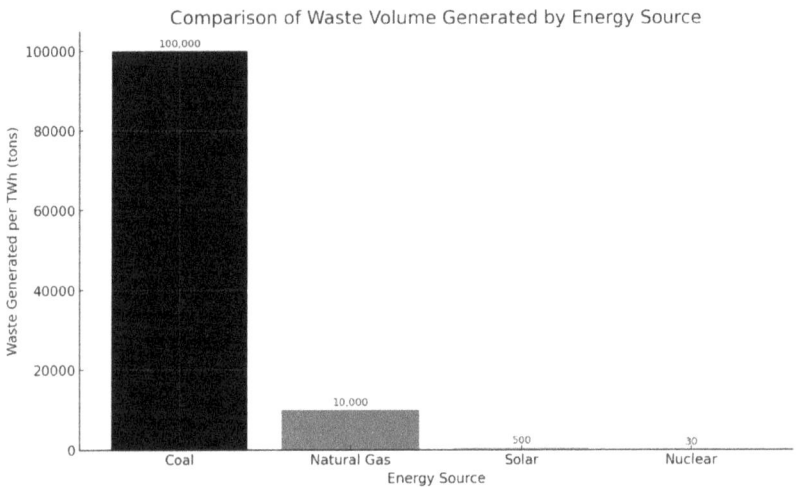

As seen, nuclear energy generates only a tiny fraction of waste compared to coal or gas, even when considering high-level waste.

"Nuclear Energy is Expensive and Slow" – Myth or Reality?

This argument has almost become a mantra in discussions about the energy transition:

"Nuclear energy is too expensive and takes decades to build. It's not worth it."

But is that really the case? The truth, as almost always, is more complex than this ready-made phrase. And when we analyze the real numbers, we realize that this argument is another misconception based on generalizations, technological prejudice, and omission of context.

Comparing the Cost of Nuclear Energy Requires Intellectual Honesty

Comparing the cost of nuclear energy with other sources is not as simple as just looking at the construction value of a plant. It is necessary to consider:

- Levelized Cost of Electricity (LCOE) over the lifecycle;
- Capacity factor (i.e., how much time the source actually generates energy);

- Lifetime duration of the installation;

- Indirect costs, such as storage, intermittency, and backup for renewables;

- Avoided cost of carbon emissions (very important for climate policies).

Table 38: Main Anti-Nuclear Myths and Scientific Rebuttals

Myth	Evidence-Based Rebuttal
Nuclear energy is the most dangerous.	Studies show nuclear has one of the lowest death rates per TWh.
Nuclear waste is unsolvable.	Long-term solutions like Onkalo (Finland) and transmutation technologies already exist.
Chernobyl killed thousands.	Most scientific estimates point to a few dozen direct deaths.
Fukushima caused a radiation disaster.	No radiation deaths occurred; impacts were mainly social and economic.
Radiation is always deadly.	All humans live with natural radiation – risk depends on dose.

Source: Own elaboration based on the data presented in the Summary Table at the end of this chapter

Levelized Cost of Electricity (LCOE)

LCOE is a standard measure used to compare the real cost of energy generation over the lifetime of different technologies.

According to the International Energy Agency (IEA) and Lazard (2023):

Table 39: Table Comparing the Levelized Cost of Electricity (LCOE)

Energy Source	LCOE (USD/MWh)
Coal	60–140
Natural Gas (Combined Cycle)	45–90
Solar Photovoltaic	35–60
Onshore Wind	30–70
Nuclear (Existing Reactors)	30–50
Nuclear (New Projects)	80–120

Source: Own elaboration based on the data presented in the Summary Table at the end of this chapter

Conclusion:

• Nuclear energy is already one of the cheapest when operating, especially in existing reactors.

• The construction costs of new projects tend to be high, mainly due to regulatory delays, bureaucracy, and lack of standardization — and not because of technical infeasibility.

• Unlike renewables, nuclear energy does not require constant backup or expensive storage systems.

Construction Speed: Delay or Planning?

Another recurring argument is that nuclear energy takes too long to be built. However, this also depends on political context and technical capacity.

Table 40: Construction Time of Nuclear Power Plants

Project	Country	Construction Time
Barakah 1	United Arab Emirates	7 years
Hinkley Point C	United Kingdom	10–12 years
Taishan 1	China	8 years
Olkiluoto 3	Finland	17 years

Source: Own elaboration based on the data presented in the Summary Table at the end of this chapter

In countries with clear political decisions and efficient regulations, it is perfectly possible to build nuclear power plants in less than 10 years.

Let us now observe this graph that compares LCOE across energy sources to better visualize the position of nuclear energy in the current scenario.

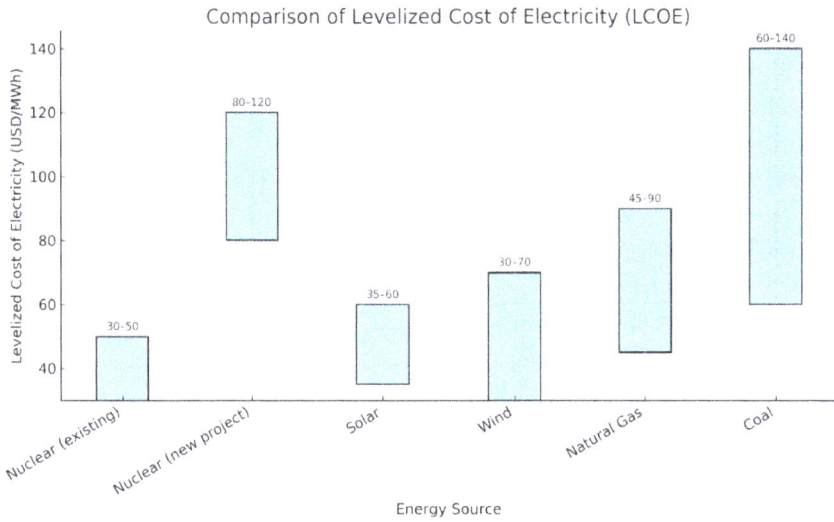

Comparison of Levelized Cost of Electricity (LCOE)

Source: Own elaboration based on the data presented in the Summary Table at the end of this chapter

Graph comparing LCOE (Levelized Cost of Electricity) between different energy sources. It clearly shows that existing nuclear energy is highly competitive, and even new nuclear projects remain within a reasonable range, especially considering the stability and longevity of production.

"Renewable Energy is Already Sufficient" – A Dangerous Myth

This is perhaps the most politically popular argument and, at the same time, technically inaccurate:

"We already have solar and wind. We don't need nuclear energy."

At first glance, this seems logical: if we can produce clean energy from the sun and wind, why continue to invest in a technology that involves radioactivity and requires heavy investment?

The answer lies in the physical reality of the electrical system and the intermittent nature of renewable sources.

Renewable Energies are Essential... But Insufficient

There is no doubt that solar and wind play a vital role in the energy transition. They are clean, abundant, and increasingly cheaper. But... they are intermittent.

- The sun does not shine at night.

- The wind does not blow every day.

- Electrical grids require stability and predictability.

The capacity factor of renewables is low:

Table 41: Capacity Factor by Energy Source

Energy Source	Capacity Factor (%)
Nuclear Energy	90–95%
Coal	60–70%
Natural Gas	50–60%
Solar Photovoltaic	10–25%
Onshore Wind	25–40%

Source: Own elaboration based on the data presented in the Summary Table at the end of this chapter

This means that we need a backup — usually fossil — to compensate for intermittent production. Alternatively, we require large-scale energy storage, which remains technically infeasible or economically inaccessible in many countries.

Let us pay attention to the following graph that compares the installed capacity of different energy sources with the energy actually delivered to the system (based on the average capacity factor). This illustrates the limitations of renewable energy without the complement of firm sources, such as nuclear.

Chart 44: Installed Capacity vs Effective Energy Delivered to the System

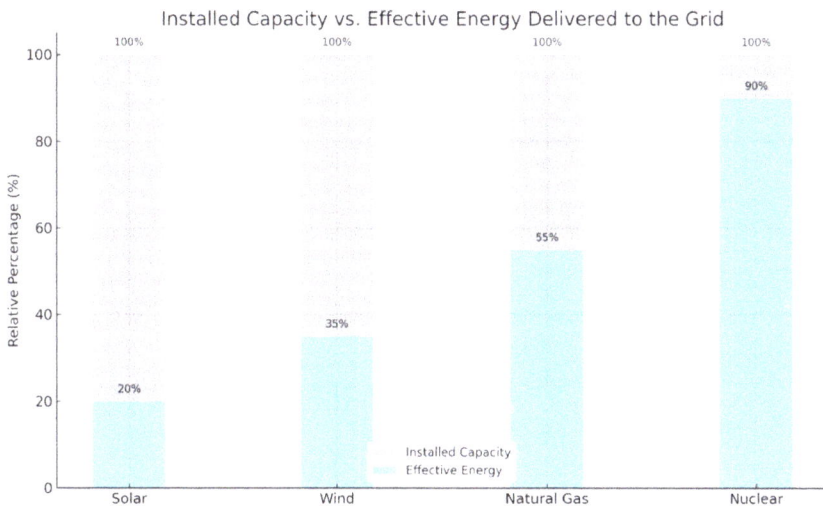

Source: Own elaboration based on the data presented in the Summary Table at the end of this chapter

Graph comparing installed capacity with the energy actually delivered to the system by different sources, i.e., the so-called "efficiency" of each energy source. It clearly shows that, although solar and wind have great potential, their actual contribution is much lower than that of nuclear energy when considering continuous and stable production.

"Nuclear Disasters Make Energy Unviable" – The Reality of Accidents

Few words evoke such emotional reactions as "nuclear accident." The mere mention of names like Chernobyl or Fukushima is enough to evoke images of tragedy, radiation, and environmental collapse.

This is one of the most used arguments by opponents of nuclear energy:

"One accident is enough to contaminate the planet. It's not worth the risk."

However, this view ignores three essential facts:

- Nuclear accidents are extremely rare.

- The number of direct victims is low compared to other industrial disasters.

- Each accident led to technological advances that made nuclear energy even safer.

Let's Look at the Facts:

Chernobyl (1986):

- It was the worst nuclear accident in history, caused by a poorly designed reactor (RBMK) without containment and operated negligently.

- Consequence: approximately 4,000 estimated deaths due to long-term effects (WHO).

- Lessons learned: the end of the use of RBMK reactors outside of Russia, global strengthening of safety standards, and the creation of the modern IAEA.

Fukushima (2011):

- It occurred after a historic tsunami, which affected the cooling system.

- Deaths from radiation: 0

- Deaths from disorganized evacuation: ~1,600 (according to the Japanese government)

- Lessons learned: Generation III+ reactors are designed to withstand external failures; passive cooling systems were implemented.

Three Mile Island (1979):

- Partial core accident, no victims, no significant external contamination.

- Lessons learned: global change in operation and monitoring protocols.

Let's visually recall the comparison between major industrial disasters (chemical, oil, and nuclear) to show the real human and environmental impact of each.

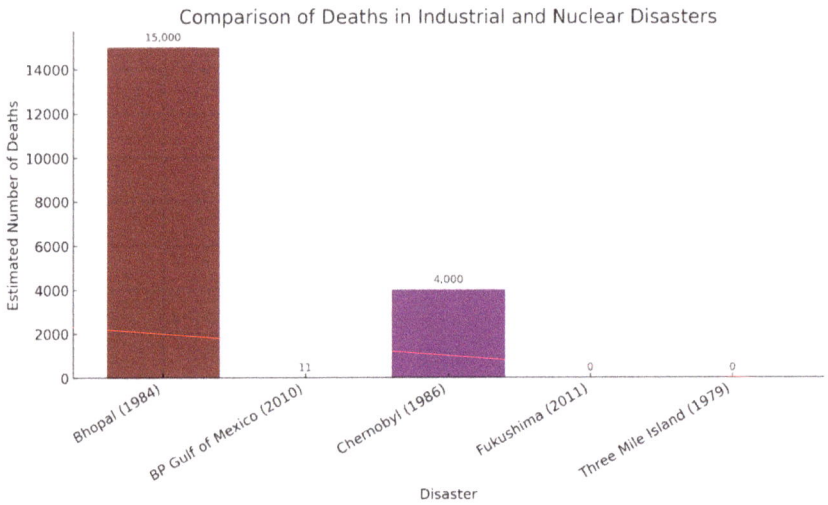

Comparison of Deaths in Industrial and Nuclear Disasters

Source: Own elaboration based on the data presented in the Summary Table at the end of this chapter

Graph comparing the estimated number of deaths in industrial and nuclear disasters, reinforcing those nuclear accidents, although media-impactful, have a much lower human impact than other industrial disasters.

Chart 46: Comparison of the Environmental Impact of Major Disasters

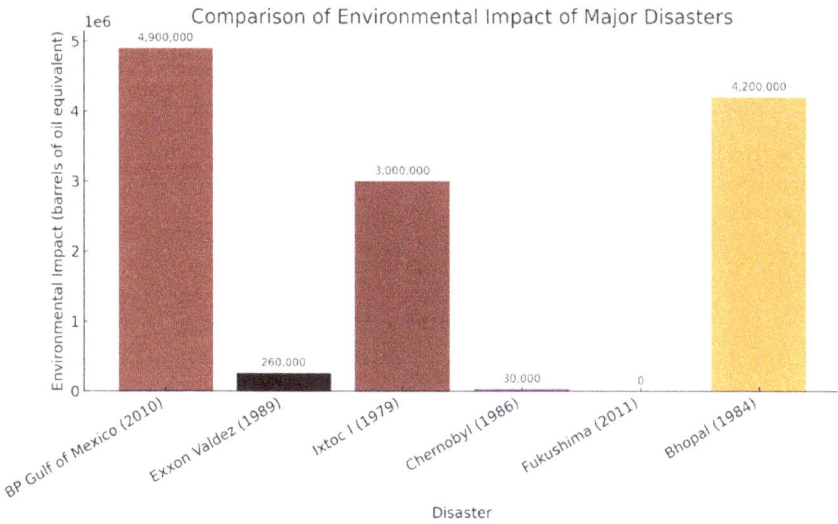

Comparison of Environmental Impact of Major Disasters

Source: Own elaboration based on the data presented in the Summary Table at the end of this chapter

Chart comparing the environmental impact (in equivalent barrels of oil) between major industrial, petroleum, and nuclear disasters.

Who Funds the Anti-Nuclear Movement?

Opposition to nuclear energy is often portrayed as a spontaneous and morally legitimate movement formed by citizens concerned about the environment. Although there are groups genuinely motivated by ethical and environmental concerns, history shows that there are much deeper and more complex interests behind the anti-nuclear movement.

In this section, we will reveal who profits from the fear of nuclear energy and how this fear has been fueled, funded, and instrumentalized over the last few decades.

The Fossil Fuel Industry – The Hidden Enemy

Nuclear energy is the only firm and large-scale source capable of replacing coal, oil, and natural gas as a primary energy source. Therefore, it represents a direct threat to the economic model of companies and countries dependent on fossil exploitation, which profit from the energy dependence of entire nations.

Historical examples:

- In the 1970s and 1980s, oil lobbies funded environmental campaigns against nuclear energy to protect their export markets.

- Recently, investigations in the US and Europe revealed that groups connected to fossil interests (especially natural gas from Russia and its allies) indirectly supported anti-nuclear campaigns through environmental NGOs.

The strategy: Promote solar and wind as the ideal solution, knowing that these technologies still require backup — usually provided by natural gas, oil, or coal, which means impoverishing those countries and maintaining their energy dependence.

Environmental Organizations – A Contradictory Relationship

Many renowned environmental NGOs (such as Greenpeace and Friends of the Earth) have a radically anti-nuclear stance, even though nuclear energy:

- Has very low CO_2 emissions;

- Has a lower environmental impact than hydroelectric and fossil fuels;

- Can safely replace polluting technologies.

The problem: These organizations receive private donations, government funds, and grants from philanthropic foundations with specific ideological or economic interests.

Known example: The Rockefeller Brothers Fund, historically involved in funding environmental campaigns, also has interests in fossil fuels and commercial renewables.

Governments and Geopolitics – Strategic Dependence

Many countries that export oil, natural gas, or coal (such as Russia, Saudi Arabia, Iran, Venezuela, etc.) have an interest in blocking the advancement of nuclear energy in other countries, as this would mean:

- Reduced fossil fuel imports;

- Less energy dependence on their clients;

- Greater technological autonomy for Western countries.

There are documented suspicions that anti-nuclear campaigns in European countries were directly supported by Russian interests, especially after the construction of the **Nord Stream**[3], aiming to increase dependence on natural gas.

---

[3] Nord Stream is an underwater gas pipeline system built to transport natural gas from Russia directly to Germany, through the Baltic Sea. The project aims to provide a direct and efficient energy supply route to Western Europe, bypassing transit countries such as Ukraine and Poland.

Nord Stream 1:

Inaugurated in 2011, with an annual capacity of about 55 billion cubic meters of gas.

Nord Stream 2:

Completed in 2021, with the same capacity, but never entered into commercial operation, due to geopolitical tensions and international sanctions.

The aim was to secure the supply of Russian gas to Europe with less political interference and lower logistical disruption risks. It became a symbol of Europe's energy dependence on Russia.

It gained prominence after the invasion of Ukraine in 2022, leading to the suspension of Nord Stream 2 and severe sanctions.

In September 2022, parts of the pipeline were damaged by mysterious explosions, sparking accusations and international investigations.

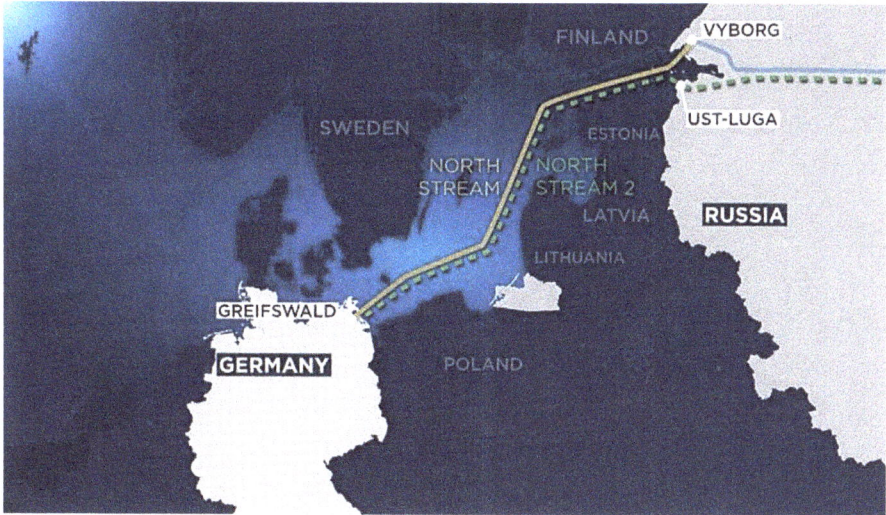

The Intermittent Energy Industry

The uncontrolled expansion of solar and wind energy has created a new multi-billion-dollar sector that relies on public incentives, subsidies, and favorable regulations.

These companies have an interest in weakening or blocking nuclear projects, which undermine the profitability of renewables when there is no demand for backup.

In countries like Germany, solar industry associations were key allies in the campaign for nuclear plant shutdowns.

The Media – The Power of Narrative

Finally, the media plays a central role in building nuclear fear.

Nuclear accidents, even when without victims, receive international and alarmist coverage, while disasters with oil, gas, or coal, even lethal, go unnoticed or are minimized.

Much of the media is funded by groups with energy interests or aligned with anti-industrial ideological views.

It is now known with the Trump Administration, through the Department of Government Efficiency (DOGE), the billions of USD that these types of organizations received through the so-called "deep state," i.e., the "hidden state" that, in exchange for obscure and secret interests, created narratives and even sometimes violent agitation to defend unconfessed interests. Nuclear energy was one of the themes to where these hidden and illegal funds have been applied. The creation of narratives was achieved by the "buying" of journalists and media outlets to spread false ideas about nuclear energy. This narrative, in turn, gave traction to radical far-left groups also funded by these funds, who placed all the irrationality in the "fight" for a "more sustainable" planet, which the media widely covered as "spontaneous" manifestations of noble citizens concerned about the planet. It is all a scam to deceive the public and make people pay for energy much more expensive than it should be if countries had an efficient energy matrix.

With the Trump Administration, I am certain that this illegal funding will end, and according to the President's own words, this term will have energy as a central point, and certainly, nuclear will be part of the attention it rightfully deserves.

Let us observe this visual infographic that represents the main interest groups that fund or promote opposition to nuclear energy, with arrows indicating their motivations.

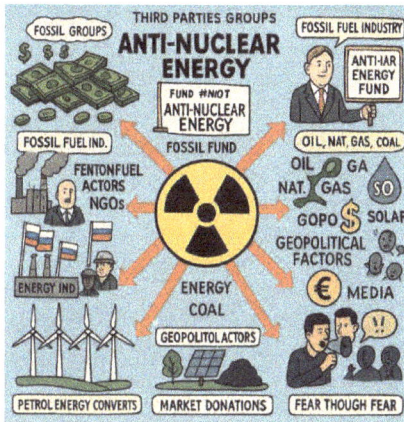

An infographic showing the main groups that fund or promote opposition to nuclear energy — with their respective motivations and relationships of influence.

Table 42: Entities and Groups Supporting Anti-Nuclear Movements

Group/Sector	Motivations	Examples of Organizations	Notes / Evidence
Fossil Fuel Industry	Protect coal, oil, and gas markets from nuclear competition.	ExxonMobil, Gazprom, Koch Industries	Support for think tanks and moderate environmental campaigns excluding nuclear.
Environmental NGOs	Anti-nuclear, anti-technological, or anti-capitalist ideology.	Greenpeace, Friends of the Earth, Beyond Nuclear	Ongoing public opposition, fear campaigns, and legal blockages.

Governments with Geopolitical Interest	Maintain energy dependence on Western countries.	Russia, Iran, Venezuela	Suspicions of funding anti-nuclear campaigns and supporting European NGOs.
Renewable Industry	Maintain dominance in subsidies and prevent firm competition.	German Solar Association, WindEurope	Active lobbying in Germany against extending the lifespans of nuclear plants.
Sensationalist Media	Audience and alignment with ideological views.	RT, Al Jazeera, The Guardian (some columnists), Documentaries like Pandora's Promise (in response)	Disproportionate coverage of nuclear accidents vs. fossil or chemical accidents.

Source: Own elaboration based on the data presented in the Summary Table at the end of this chapter

Table 43: Key Interests Behind Nuclear Opposition

Group / Interest	Likely Motivation
Fossil fuel industry	Avoid competition from stable clean energy
Radical environmental groups	Ideological anti-technology or degrowth views
Governments with populist agendas	Gaining public support through symbolic decisions
Sensationalist media	Exploiting fear to drive audience engagement
Pacifist movements	Confusing civil nuclear with nuclear weapons

Source: Own elaboration based on the data presented in the Summary Table at the end of this chapter

The German Case – Closing Nuclear Plants and Increasing Emissions

Germany was, for decades, a technological leader in nuclear energy. However, after the Fukushima accident (2011), the country decided to close all its nuclear plants, citing safety concerns. This politically motivated decision, supported by environmental groups, became known as the "Energiewende" – the German energy transition.

But the reality was very different from the rhetoric. The replacement of nuclear energy did not happen through clean and renewable energy sources, as many believe. It happened mainly through... coal and natural gas.

What did Germany do?

In 2011, after Fukushima, the Merkel government decided:

- Immediately shut down eight nuclear reactors.

- Close all remaining reactors by 2023.

- Replace nuclear energy with solar, wind... and Russian natural gas.

In April 2023, Germany shut down its last three nuclear reactors — even during an energy crisis caused by the war in Ukraine and the reduction of gas supplies from Russia.

The result: Increased Emissions and Higher Electricity Bills.

Table 44: Impact of Nuclear Shutdown in Germany

Indicator	Before Energiewende (2010)	After Nuclear Shutdown (2023)
Nuclear Share in the Matrix (%)	22%	0%
Coal Share (%)	28%	31%
Russian Gas Imports (%)	37%	0% (replaced by LNG)
Electricity Price (€ MWh)	~50	>150
CO_2 Emissions (Mt/year)	~760	~810

Source: Own elaboration based on the data presented in the Summary Table at the end of this chapter

Conclusion: Closing nuclear power plants has increased emissions, external dependence, and costs for consumers.

This graph will visually show how the reduction in nuclear energy has been compensated by fossil fuels and not by clean sources.

Chart 47: Evolution of Electricity Production in Germany by Source

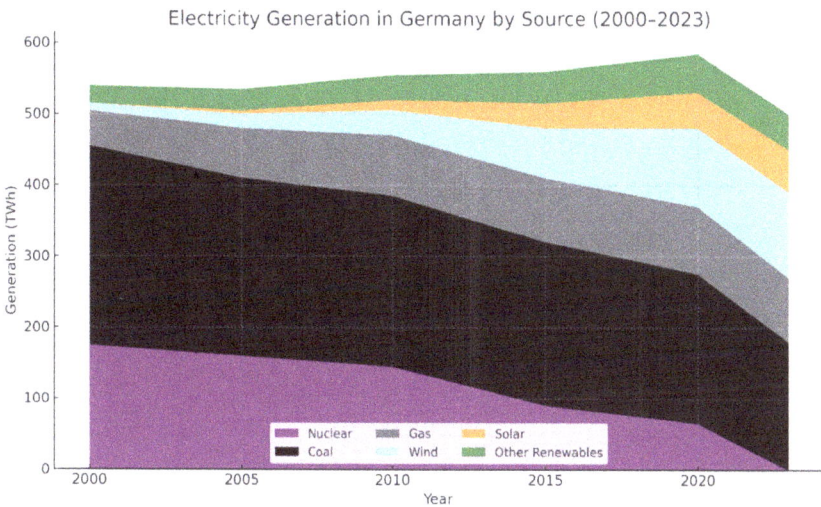

Electricity Generation in Germany by Source (2000-2023)

Source: Own elaboration based on the data presented in the Summary Table at the end of this chapter

Graph of the evolution of electricity generation in Germany by source (2000–2023). It clearly shows how the reduction of nuclear energy has been compensated by coal and gas, with renewables growing but not fully replacing nuclear power.

Angela Merkel and the End of Nuclear Energy in Germany: When Politics Ignores Science

Angela Merkel is often portrayed as one of the most influential leaders of the 21st century. A scientist by training with a PhD in physics, she was Chancellor of Germany from 2005 to 2021 after German reunification and was seen by many as a symbol of stability and European pragmatism.

But one chapter of her leadership remains highly controversial: As Chancellor (2005-2021), she led Europe with international prestige, a Europe in need of leadership and direction, but the decision to destroy Germany's nuclear program, made under popular and ideological pressure after the Fukushima accident in 2011, was emotional and political, not scientific. This choice would radically change Germany's energy matrix and bring profound consequences for its economy, security, and geopolitical autonomy.

Timeline: Merkel's Energy Transition

- Before 2011: Germany had 17 active nuclear reactors, which generated over 22% of the national electricity at low cost and with extremely low CO_2 emissions.

- March 2011: A tsunami in Japan hits Fukushima. No direct relation to Germany.

- April 2011: Merkel succumbs to pressure from public opinion and environmental NGOs.

- The government decides to shut down eight reactors immediately and decommission the rest by 2022.

- The "Energiewende" program begins, based on solar, wind... and Russian gas.

Table 45: Timeline for the Shutdown of All Nuclear Plants in Germany

Year	Political Milestone	Consequence
2005	Merkel becomes chancellor	Promises to modernize and decarbonize Germany's energy mix
2010	Nuclear = 22% of electricity mix	Germany is a reference for clean and reliable energy
2011	Fukushima accident in Japan	Merkel decides to shut down nuclear reactors in Germany
2012	Progressive shutdown begins	Russian gas and coal begin replacing nuclear generation
2020	Nuclear nearly phased out	Dependence on Russian gas exceeds 50%
2022	Invasion of Ukraine	Energy crisis with price surge
2023	The last reactors shut down	German industry was impacted, and companies relocated to China and USA

Source: Own elaboration based on the data presented in the Summary Table at the end of this chapter

A Scientist Who Ignored Science?

Paradoxically, Merkel, with a scientific background and deep technical knowledge, made a political decision based on fear and emotion, not evidence.

Instead of rationally assessing the safety of German reactors – considered among the safest in the world – she opted for a symbolic and ideological gesture, applauded internationally, but with incalculable costs.

TIMELINE – ANGELA MERKEL'S DECISIONS ON NUCLEAR ENERGY

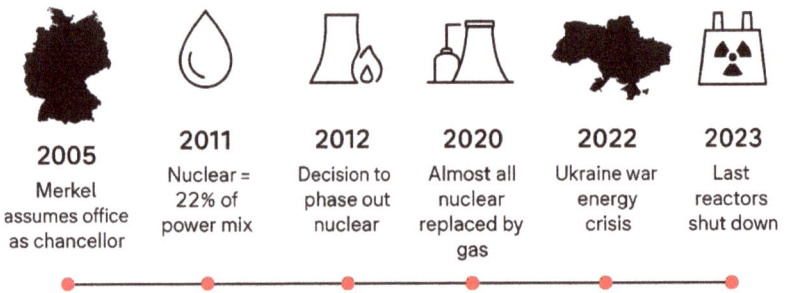

2005	2011	2012	2020	2022	2023
Merkel assumes office as chancellor	Nuclear = 22% of power mix	Decision to phase out nuclear	Almost all nuclear replaced by gas	Ukraine war energy crisis	Last reactors shut down

The Consequences: More Emissions, More Dependence, Less Industry

- Electricity generated by nuclear plants was primarily replaced by coal and Russian natural gas.

- Germany's energy dependence on Russia rose to over 50% of the gas supply before the war in Ukraine.

- With the closure of Nord Stream, energy prices skyrocketed (see graph), reaching over 150 €/MWh in 2023.

- The German industrial sector – especially chemical, steel, and automotive – was forced to reduce production or move operations to countries like China, the US, and Norway.

Result: The strongest economy in Europe placed its energy matrix in the hands of an autocratic regime and lost global competitiveness.

Chart 48: Evolution of Electricity Price in Germany

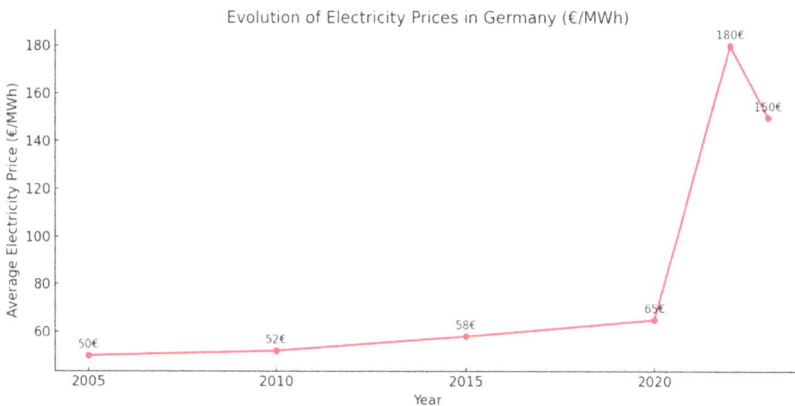

Evolution of Electricity Prices in Germany (€/MWh)

Source: Own elaboration based on the data presented in the Summary Table at the end of this chapter

Graph of the evolution of electricity price in Germany (€/MWh). It clearly shows the exponential increase after the closure of the nuclear program and the dependence on Russian gas.

Merkel in Her Own Words... and the Facts:

"Fukushima changed my view on nuclear energy." – Angela Merkel, 2011

But... Fukushima killed zero people from radiation. And the German reactors have no technical resemblance to those in Japan.

Lesson for the Future: When Politics Ignores Science, Everyone Pays the Price.

This case clearly shows that:

- Fear can be a terrible advisor for strategic decisions.

- A nation's energy stability cannot depend on ideologies or populist pressures.

- Nuclear energy was sacrificed in Germany for political reasons, and the country is paying a very high price.

Table 46: Contradictions in Energy Policies

Declared Policy	Actual Action	Consequence
Rapid decarbonization	Closing safe nuclear plants	Increased use of gas or coal
Reduce foreign dependency	Importing energy instead of producing nuclear	Loss of energy sovereignty
Support science and innovation	Ignoring next-generation reactor advancements	Technological stagnation
Ensure energy security	Eliminating stable nuclear source	Intermittency and blackout risks

France and Finland: The Choice of Reason, Not Fear

While Germany gave in to fear and dismantled its nuclear sector, France and Finland took the opposite path. They chose to strengthen, modernize, and expand their nuclear capabilities, recognizing that there is no viable energy transition without firm, clean, and reliable sources.

This difference in strategy became especially evident during the European energy crisis of 2022, caused by the war in Ukraine and dependence on Russian gas.

France: The European Nuclear Pioneer

As we have seen earlier, France has been a world leader in civilian nuclear energy since the 1970s, driven by the desire for energy self-sufficiency after the oil crisis.

Key data:

- It has 56 active nuclear reactors.

- More than 70% of its electricity comes from nuclear energy – the highest proportion in the world.

- It emits less CO_2 per capita in electricity generation than almost all industrialized countries.

- It exports electricity to neighboring countries (including Germany itself...).

In 2022, President Emmanuel Macron announced a plan to build 6 new EPR reactors and keep existing ones in operation and safe for several decades.

"Without nuclear, there will be no European energy sovereignty." – Macron

Finland: A Small Country with a Vision for the Future

Finland decided that the most rational and secure path to decarbonize its electricity matrix was to firmly invest in nuclear.

Highlights:

- It operates five nuclear reactors, which provide over 35% of the country's electricity.

- In 2023, the Olkiluoto-3 reactor, the largest and most powerful in Europe (EPR – 1,600 MW), came into operation.

- Finland is the first country in the world to complete a final geological repository for nuclear waste (Onkalo).

- The population largely supports nuclear energy, with over 70% acceptance.

Table 47: Energy Comparison: Germany vs. France vs. Finland

Indicator	Germany	France	Finland
% of electricity from nuclear	0%	70%	35%
CO_2 emissions per capita (electric)	High (~8.5 t)	Low (~2.5 t)	Very low (~1.8 t)

Gas dependency	High (~80%)	Moderate (~30%)	Low (~20%)
Average electricity price (€)	>150 €/MWh	~85 €/MWh	~65 €/MWh
Public support for nuclear energy	<40%	>60%	>70%

Source: Own elaboration based on the data presented in the Summary Table at the end of this chapter

Chart 49: Energy Indicators: Germany vs. France vs Finland

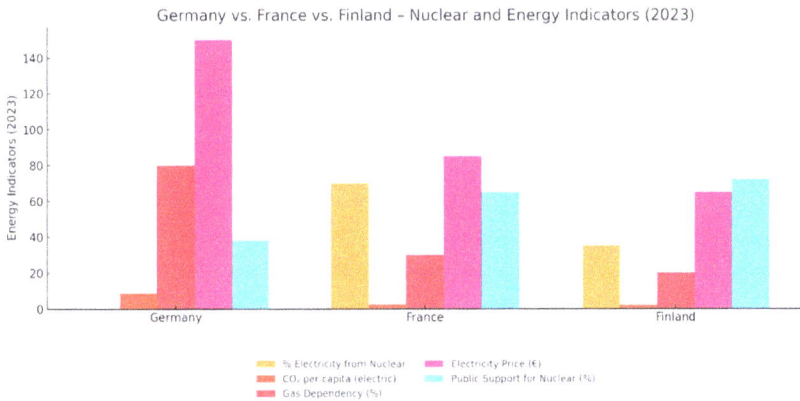

Source: Own elaboration based on the data presented in the Summary Table at the end of this chapter

The comparative graph between Germany, France, and Finland highlights key nuclear and energy indicators in 2023.

Conclusion of the Present Chapter - Between Fear and Reason – The Future of Energy Is at Stake

The history of nuclear energy in the past decades is not just a story of science, engineering, and energy policy. It is, above all, a story about how fear can silence knowledge, how ideologies can eclipse reason, and how ill-founded decisions can have profound, lasting, and global consequences.

In this chapter, we have seen how opposition to nuclear energy often does not stem from technical reality but from the construction of narratives — narratives fueled by economic interests, ideological pressures, disinformation, and sometimes pure ignorance.

We have seen how Angela Merkel, a respected leader and scientist by training, abandoned scientific logic in favor of political symbolism — with disastrous consequences for Germany and all of Europe.

We contrasted this path with that of France and Finland, which chose to invest in nuclear energy with clarity, long-term vision, and responsibility toward the environment and future generations.

And now, the question turns to the reader:

What path is your country taking?

Is it giving in to fear and public pressure, or is it betting on an energy transition based on science, safety, and stability?

Nuclear energy is not perfect — no energy source is.

But it is the only one capable of generating electricity on a large scale, with very low emissions, 24 hours a day, without relying on wind, sun, or fossil fuels.

In a world in climate and geopolitical crisis, refusing nuclear energy for ideological reasons is a luxury humanity can no longer afford.

Table 48: Summary Table: Key Lessons on Nuclear Energy and Global Opposition

Lesson	Reflection / Implication
Opposition to nuclear energy is often not technical but ideological or strategic.	It is essential to investigate who funds the anti-nuclear narrative and why.
Political decisions can destroy decades of energy progress.	Germany's case illustrates the severe consequences of fear-based choices.
Nuclear energy is one of the cleanest and safest methods for producing electricity at scale.	It should be an essential part of any realistic energy transition strategy.
Countries investing in nuclear energy have cheaper electricity, lower emissions, and greater security.	Examples from France and Finland highlight the benefits of data-driven decisions.
The public needs access to clear and objective information about energy.	This book aims to be a tool for encouraging more rational and informed discussion.

Source: Author's own elaboration based on various studies and media reflections.

Table 49: Sources Consulted in Chapter 6

Source	Description	Notes
Our World in Data (2022)	Comparative study of mortality per TWh of energy generated.	Used for the comparison of mortality between energy sources.
International Energy Agency (IEA)	Reports on Levelized Cost of Electricity (LCOE) and nuclear energy.	Relevant source for cost and efficiency data of energy sources.
Lazard (2023)	Study on Levelized Cost of Electricity (LCOE).	Used to compare nuclear competitiveness against renewables and fossil sources.
Greenpeace	Environmental NGO with an anti-nuclear position.	Mentioned as an example of organizations advocating for renewable energies over nuclear.
Friends of the Earth	Anti-nuclear environmental NGO.	Represents one of the groups funding anti-nuclear opposition.
Nord Stream	Information about the Nord Stream pipeline and its impact on European energy.	Used to explain the energy policy of Germany and its dependence on Russian gas.

Rockefeller Brothers Fund	Foundation supporting environmental NGOs and renewables.	Related to the funding of anti-nuclear campaigns.
Merkel, Angela (2011)	Merkel's statements on changing stance on nuclear energy after Fukushima.	Direct source of Merkel's declarations about German nuclear policy.

Preparation for the Next Chapter: Nuclear Geopolitics: Energy, Power, and Global Influence

Throughout this chapter, we have uncovered the arguments behind opposition to nuclear energy and shown how many of them do not withstand critical, technical, and factual analysis. We have understood that sometimes the most significant decisions are not made based on science or reason but under the pressure of political, ideological, or geostrategic narratives.

But if fear and misinformation are driving forces behind the opposition, power and influence are often the real foundations of nuclear energy.

That is why, in the next chapter, we will dive into an even deeper dimension: **the geopolitics of nuclear energy.**

We will explore how access to nuclear technology shapes international alliances, how major powers use nuclear energy as an instrument of power, and why some countries pursue it with such determination — while others reject it at all costs.

Because it is more than just an energy source, nuclear power is also a tool of sovereignty, a symbol of prestige, and a strategic trump card in the great game of nations.

Chapter 7 –
Nuclear Geopolitics:
Energy, Power, and Global Influence

Throughout history, energy has always been synonymous with power. From the mastery of fire to the rise of major industrial powers driven by coal and oil, access to secure, abundant, and controllable energy sources has shaped empires, fueled wars, and defined the destinies of nations.

In the 21st century, this reality remains — but with a crucial nuance: energy is no longer just a matter of resources but of strategy, sovereignty, and global influence. And at the center of this new geopolitical game stands a key element: **nuclear energy**.

More than just a technology for generating electricity, nuclear power represents:

- A symbol of scientific and industrial autonomy.

- A diplomatic and military asset at international negotiation tables.

- And, in many cases, a dividing line between regional powers and major global powers.

Today, countries that master the nuclear fuel cycle, export reactors, or control supply chains exert political influence far beyond their energy borders.

At the same time, misinformation, fear, and ideological opposition have been used as geostrategic instruments — hindering the advancement of nuclear energy in countries that could otherwise become energetically independent and politically stronger.

This chapter explores nuclear energy as a central element of modern geopolitics. We will examine:

- Who holds nuclear power and how it is used.

- How access (or lack thereof) to nuclear technology shapes international relations.

- And why energy — especially nuclear — will be one of the main axes of influence, competition, and sovereignty in the 21st century.

In this new global chessboard, those who control nuclear energy not only generate electricity — they control the game itself.

Between the Atom and Sovereignty

Since the United States detonated the first nuclear bomb in July 1945, the world entered a new geopolitical era. For the first time in human history, a single country held a destructive power so colossal that no other nation could match it. It was the birth of the nuclear age — and with it, the clear separation between those who control the core of the atom and those who depend on those who do.

Civil nuclear energy emerged almost simultaneously as an ethical and technological counterpoint to atomic weapons, promising nearly unlimited, clean, and sovereign electricity. However, this peaceful aspect of the technology has never been entirely separate from its strategic potential. In fact, the mere fact that a country masters the complete cycle of nuclear technology — even for civilian purposes — is enough to alter its position on the global chessboard.

Scientific Sovereignty as a Tool of Prestige

Mastering nuclear energy is not just a technical achievement — it is an unequivocal sign of scientific capability, institutional maturity, and industrial autonomy. It is no coincidence that:

- Only a small group of countries possesses the technology to enrich uranium, build reactors, and handle nuclear waste.

- These countries are seen as 'first-tier' nations, even if they do not possess nuclear weapons.

In international diplomacy, nuclear knowledge equates to negotiating power:

- It provides leverage in treaties.

- It guarantees respect in multilateral forums.

- It prevents foreign interference in strategic and energy decisions.

Deterrence: Reality or Potential?

Even without explicit military intentions, merely mastering the technology creates what is known as 'latent deterrence':

- A country with a robust civil nuclear infrastructure could, in theory, rapidly convert that know-how into a military program.

- This implicit possibility makes such countries much harder to intimidate, sanction, or isolate.

Japan is a prime example:

- It has never developed nuclear weapons.

- Yet it holds dozens of tons of reprocessed plutonium.

- It possesses the technological and scientific capacity to assemble an arsenal in a short time if national security demands it.

Civil Nuclear Energy as a Pillar of Energy Sovereignty

Access to nuclear energy enables a country to:

Drastically reduces its dependence on fossil fuel imports

- Stabilize its long-term energy matrix.

- Protect itself from external geopolitical shocks (such as sanctions, wars, or trade extortion).

Thus, nuclear energy becomes an invisible but powerful shield. In times of international tension, it is often the last bastion of a State's sovereignty.

External Perception: Fear, Respect, or Alignment?

The world watches closely the countries that:

- Build nuclear power plants with their own technology.
- Enrich uranium within their borders.
- Develop reprocessing or recycling technologies.

Often, this observation is accompanied by diplomatic pressure, accusations of militarization, or attempts to limit scientific advancement under the pretext of global security.

At its core, there is a systemic fear of countries that become self-sufficient in nuclear energy, as it also means political, economic, and strategic independence.

The atom is simultaneously a source of light — and a shadow of power.

The Concept of Deterrence – Between Fear and Strategic Balance

Since the Cold War, the logic of nuclear deterrence has been the foundation of strategic stability among armed powers.

The doctrine of Mutual Assured Destruction (MAD) establishes that if two countries possess enough nuclear weapons to destroy each other, neither will dare initiate a direct conflict — as the cost would be their own annihilation.

This balance of terror, paradoxical as it may seem, prevented direct wars between major powers for more than half a century, even during moments of high tension (such as the Cuban Missile Crisis in 1962).

Explicit Deterrence: The Armed Club

The countries officially possessing nuclear weapons — the United States, Russia, China, France, the United Kingdom, India, Pakistan, North Korea (and presumably Israel) — use this capacity as an absolute shield of sovereignty.

Possession of nuclear warheads:

- Discourages invasions, military pressures, and external extortion.
- Elevates the country to a superior geopolitical status.
- Ensures a privileged seat in international decision-making.

None of these countries has been invaded or subjected to regime change imposed by external forces, largely because the nuclear risk acts as an insurmountable red line.

Latent Deterrence: The Power of Those Who Can, Even If They Do Not Intend To

There is also a subtler — and no less effective — type of deterrence.

Even without warheads or military tests, a country with complete mastery of the civil nuclear cycle can become 'unreachable' by external pressure.

This phenomenon is known as 'latent deterrence':

The technical and industrial capacity to produce nuclear weapons if the security context requires it — even if there is no declared intention to do so.

Examples of Latent Deterrence in Practice:

Japan:

- Holds over 45 tons of stored (reprocessed) plutonium, enough for thousands of warheads.

- Possesses one of the most advanced nuclear sectors in the world.

- Despite its pacifist Constitution, it is widely recognized as a 'latent nuclear power.'

- In the event of a collapse of the alliance with the U.S. or a severe regional threat (e.g., from North Korea or China), it could assemble an arsenal within months.

Germany:

- Despite abandoning civil nuclear energy, it maintains enormous scientific and industrial capacity.

- Participates in NATO's 'nuclear sharing' agreements, with technical and logistical access to U.S. armament.

- Plays a central role in international nuclear negotiations, even without possessing weapons.

Brazil:

- Has an independent nuclear program, including the only naval reactor under construction in Latin America.

- Achieves 100% national uranium enrichment.

- Has never had weapons but is considered a full strategic potential State.

- Article 4 of the Brazilian Constitution allows revising the peaceful policy in case of threats to sovereignty.

South Korea:

- Highly technologically advanced.

- Access to American and Japanese nuclear technology.

- The growing threat from North Korea generates internal pressure for strategic rearmament.

Table 50: Nuclear Deterrence: Explicit vs Latent

Type of Deterrence	Countries	Characteristics
Explicit Deterrence	United States, Russia, China, France, United Kingdom, India, Pakistan, North Korea, Israel (presumed)	Possess declared or operational nuclear arsenals; use nuclear capability as a military shield and symbol of geopolitical status.
Latent Deterrence	Japan, Germany, Brazil, South Korea, Canada	Scientific and technical mastery of the nuclear cycle; civil-military conversion capability; diplomatic prestige without possessing warheads.

Source: Own elaboration based on the data presented in the Summary Table at the end of this chapter

Geostrategic Respect

These countries are often treated with the same diplomatic caution as nations with declared nuclear arsenals because:

- They cannot be easily intimidated.

- They possess technological and economic retaliation capabilities.

- They participate in global negotiations with greater autonomy.

- This is the 'hard-soft power' of nuclear deterrence:

It is not about threatening the world with destruction but about placing oneself beyond the reach of geopolitical submission.

Nuclear deterrence, whether explicit or latent, remains one of the most powerful instruments of strategic stability — and also of geopolitical inequality.

In the realm of powers, nuclear capability continues to be the ultimate line of defense of sovereignty.

And the mere fact of being able to possess it... is often enough to deter any risk.

Perception of Asymmetries in the International System

Although the global nuclear control regime, primarily based on the Treaty on the Non-Proliferation of Nuclear Weapons (NPT) and supervised by the International Atomic Energy Agency (IAEA), has helped to prevent the uncontrolled proliferation of nuclear weapons, it is not without criticisms regarding its unequal application.

These criticisms come not only from so-called 'revisionist' or contesting States but also from democratic nations committed to the peaceful use of atomic energy and to the sovereign right to technological development.

Table 51: Non-Permanent Countries with Nuclear Capabilities or Ambitions

Country	Nuclear Status	Remarks
India	Declared nuclear power	Not a signatory to the NPT; tests conducted in 1974 and 1998
Pakistan	Declared nuclear power	Developed weapons in response to India; not a signatory to the NPT
North Korea	Declared nuclear power	Withdrew from the NPT; conducted several tests
Israel	Presumed nuclear capability	Neither confirms nor denies possession; has not joined the NPT
Iran	Advanced technical capability	NPT signatory; supervised by the IAEA; subject of controversy
Brazil	Advanced civil program	NPT signatory; no weapons; has full fuel cycle mastery
Japan	Full technological capability	NPT signatory; possesses large amounts of civilian plutonium

| Germany | High technical capacity | NPT signatory; participates in NATO nuclear sharing |
| South Korea | Strategic potential | NPT signatory; ongoing internal debate on armament |

Source: Own elaboration based on the data presented in the Summary Table at the end of this chapter

Table showing the non-permanent members of the Security Council that possess nuclear weapons, technical capabilities, or relevant ambitions in the nuclear field

1. The Modernization of Arsenals by Official Holders

The five permanent members of the Security Council (P5)[4], who are also the five recognized as 'nuclear-weapon states' under the NPT, continue to:

- Maintain considerable arsenals (some with thousands of active warheads).

- Invest in technological modernization, new launch platforms, and advanced simulations.

- Extend the service life of existing weapons, in contradiction to Article VI of the NPT, which calls for gradual disarmament.

This creates the perception that the major powers demand restraint from others but are unwilling to lead by example.

---

[4] USA, Russia, China, UK and France

Table 52: Countries with Advanced Military
and Civil Nuclear Capabilities

Countries with Declared or Presumed Nuclear Weapons	Countries with Advanced Civil Nuclear Programs (no arsenal)
United States	Germany
Russia	Japan
China	Brazil
France	Canada
United Kingdom	South Korea
India	Finland
Pakistan	Sweden
North Korea	Argentina
Israel (presumed)	United Arab Emirates
	Belgium
	Netherlands

Source: Own elaboration based on the data presented in the Summary Table at the end of this chapter

2. The Accepted Ambiguity of Certain States

Countries like Israel, which have never adhered to the NPT, are widely regarded as possessing nuclear weapons. However:

- They do not officially acknowledge their arsenal.

- They are not subject to regular IAEA inspections.

- They are diplomatically protected by influential allies, which limits any effective multilateral action.

This double standard, tolerated by the international system, undermines the credibility of the non-proliferation regime in the eyes of other countries, especially those in the Global South.

Nevertheless, and to be completely honest, it is important to emphasize that Israel is a democratic state governed by the rule of law, where strong institutions exist to limit or even prevent the use of its nuclear arsenal. The fact that Israel possesses a considerable estimated arsenal does not mean it will use it recklessly; rather, it serves as a deterrent to ensure its survival as a state and to send a strong warning to its adversaries.

3. The Intense Scrutiny of Certain States

In contrast, countries like Iran — a signatory of the NPT and subject to rigorous IAEA inspections — face:

- Constant diplomatic pressure.

- Severe economic sanctions, even when technically in compliance with their commitments.

- Disproportionate political reactions to any technical steps that might be interpreted as suspicious, even within legally permitted limits.

Such selective treatment creates tensions, even among countries without military intentions, but that demand respect for their sovereign technological development.

However, it is important not to forget that Iran is a religious state where ultimate power rests in the hands of radical clerics, without scrutiny from judicial powers or democratically elected bodies. Thus, the suspicions surrounding the development of its nuclear program are entirely legitimate despite the country seemingly being treated unequally.

4. The Case of Brazil: Sovereignty, Transparency, and Mistrust

Brazil is a paradigmatic example of a country that:

- Signed and ratified the NPT.

- Has a history of peaceful use of nuclear energy.

- Is one of the only countries in the world to enshrine in its Constitution the prohibition of nuclear weapons.

- Created, together with Argentina, the ABACC agency — a pioneering binational safeguards model.

Even so, the country faces:

- Resistance to accessing certain sensitive technologies, such as the closed reprocessing cycle.

- Implicit suspicion from traditional suppliers, who impose conditions not applied to closer allies.

Such blockages are often justified on technical grounds but are perceived as political obstacles to sovereign development. Unfortunately, Brazil is still perceived as a 'weak state' where institutions are permeable, thus the resistance to its nuclear development. Currently, it appears that the country is governed more by the Supreme Court than by its legally constituted powers. These factors weigh heavily against the transfer of nuclear technology due to the manifest lack of confidence in the country's institutions.

5. The North-South Imbalance and the Geopolitics of Technology

These asymmetries generate a growing sense of structural injustice within the global nuclear system:

- Global North countries tend to control strategic nuclear technology and inputs.

- Global South countries are often treated as pupils under surveillance, even when complying with all treaties.

- Access to nuclear energy is conditioned not only by technical criteria but also by political alignments and regional alliances.

What many emerging countries denounce is not control per se but selective control.

The current system allows a few to define the rules... and to alter the criteria according to geopolitical convenience.

The nuclear non-proliferation regime is, in essence, a diplomatic construction based on mutual trust, transparency, and multilateral cooperation. However, its effectiveness depends on the perception of justice and impartiality.

If the criteria appear to fluctuate, if sanctions hit some and spare others, and if access to civil nuclear technology continues to be restricted for political reasons, then the risk is the erosion of voluntary adherence to the system — and with it, the loss of its legitimacy.

The Control of Technology and Non-Proliferation Regimes

After the outbreak of the nuclear age and the multiplication of atomic weapons in the 1950s and 1960s, the international community recognized the need to control access to sensitive technologies and to prevent a widespread arms race.

Thus, the Treaty on the Non-Proliferation of Nuclear Weapons (NPT) was born, signed in 1968, and force since 1970, based on three central pillars:

1.Non-Proliferation:

Countries that already possessed nuclear weapons in 1967 (the P5) committed not to transfer nuclear weapons or military nuclear knowledge to other states.

The other signatories committed not to produce nuclear weapons under any circumstances.

2. Gradual Disarmament:

The nuclear-armed States were to negotiate in good faith measures to reduce and eventually eliminate their arsenals — a commitment that still draws criticism regarding its actual implementation.

3. Peaceful Use of Nuclear Energy:

All signatories have the right to access nuclear technology for civilian purposes (electricity generation, health, agriculture, etc.) under conditions of transparency and international inspection.

The NPT has become the legal and diplomatic pillar of the global nuclear control system, now with 191 signatory States, making it one of the most universally accepted treaties.

The Role of the IAEA – The Watchdog of the Atomic World

The International Atomic Energy Agency (IAEA), based in Vienna, is the United Nations body responsible for:

- Verifying that countries are using nuclear energy solely for peaceful purposes.

- Conducting technical inspections of nuclear facilities.

- Monitoring uranium and plutonium inventories.

- Investigating suspicions of diversion or undeclared activities.

The IAEA operates based on safeguard agreements that countries sign either voluntarily or as required under the NPT. In some cases (such as Iran), there are also additional protocols allowing for more invasive and short-notice inspections.

Its work is essential but depends on the cooperation of States and the political support of UN members.

Table 53: Pillars of the NPT and Functions of the IAEA

NPT Pillars	IAEA Functions
Non-Proliferation: Prevent the spread of nuclear weapons beyond the five recognized countries.	Inspection and verification of nuclear facilities to ensure peaceful use.

Disarmament: Commitment by nuclear-armed states to reduce their arsenals.	Monitoring of nuclear materials (uranium, plutonium).
Peaceful Use: Guarantee the right to access civil nuclear energy under international verification.	Oversight of safeguards agreements and additional protocols.

Source: Own elaboration based on the data presented in the Summary Table at the end of this chapter

Limitations and Challenges of the NPT and the IAEA

Despite their crucial role, both the NPT and the IAEA face complex geopolitical challenges:

- The NPT recognizes as 'legitimate' only the arsenals of the P5, perpetuating the imbalance.

- Countries like Israel, India, Pakistan, and North Korea are not formally bound by their obligations (either by not signing or by withdrawing from the treaty).

- The IAEA does not have the authority to punish violations — it can only report them, depending on action by the Security Council.

- Access to civil technology is, in practice, hindered by political and commercial restrictions that go beyond technical safeguards.

The Geopolitics of Isotopes: Uranium, Plutonium, and Invisible Power

Behind facilities, treaties, and inspections lies the raw material of nuclear power:

- Natural uranium is abundant, but only the isotope U-235 (less than 1%) is fissile. To be used in reactors (3–5%) or weapons (>90%), it must be enriched — a technically demanding and strategically sensitive process.

- Plutonium-239 can be generated in reactors from uranium and extracted through reprocessing, another sensitive, dual-use technology (civilian or military).

Controlling these technologies means controlling access to the frontier between energy and weaponry.

The Suppliers Club and Technological Conditioning

In addition to the NPT, there is the so-called Nuclear Suppliers Group (NSG), an informal association of countries that control the trade of nuclear materials and equipment.

This group:

- Regulates other countries' access to cutting-edge technology.

- Imposes additional conditions for exports, often based on political rather than purely technical considerations.

- Is one of the main mechanisms through which major powers limit the expansion of nuclear technology in emerging countries, even when these countries comply with the IAEA.

The international non-proliferation system is indispensable but far from perfect. Balancing the sovereign right to development with global security is a constant challenge — and, at times, a diplomatic battlefield.

Those who control uranium, plutonium, and inspection protocols ultimately control access to power.

NUCLEAR FUEL CYCLE
with principal international control points highlighted

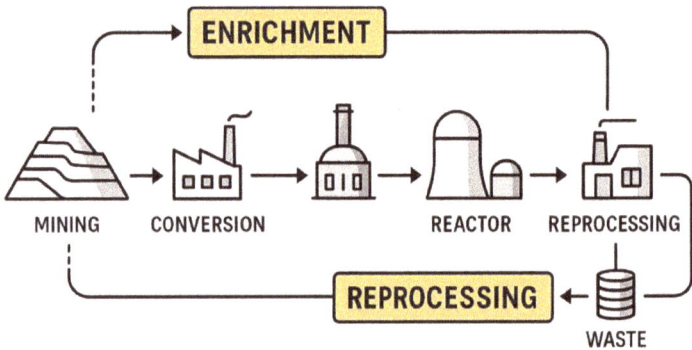

Illustrative diagram of the nuclear fuel cycle, highlighting the main points of international control (such as enrichment and reprocessing).

Access to Uranium and the Race for Supply Chains

Contrary to popular belief, nuclear power does not begin in laboratories or power plants. It begins underground, with a heavy metal of atomic number 92: uranium.

Uranium is the essential raw material for most current nuclear reactors, and its life cycle includes:

- Extraction (mining),

- Chemical conversion,

- Isotopic enrichment,

- Fuel fabrication,

- And finally, return as waste or reprocessed materials.

Controlling this chain is, therefore, a strategic imperative for any country aiming for energy autonomy based on nuclear power.

Major Uranium Producers in the World

Currently, global uranium production is concentrated in a few countries, making the supply vulnerable to political shocks, instability, and commercial manipulation.

Table 54: Main Uranium Producers (2023)

Country	Share in Global Production (%)
Kazakhstan	40%
Canada	15%
Namibia	11%
Australia	8%
Uzbekistan	6%
Niger	4%
Others	16%

Source: Own elaboration based on the data presented in the Summary Table at the end of this chapter

Although countries like the United States, Russia, and China have reserves, they heavily depend on imports to maintain their active programs.

The Geopolitics of Nuclear Supply Chains

Uranium extraction is just the beginning. The subsequent steps — conversion, enrichment, transport, and reconversion of fuel — are dominated by a restricted club of countries with technical infrastructure and robust bilateral agreements.

- Enrichment: Led by Russia (Rosatom), France (Orano), the United States (Centrus, Urenco), and China.

- Conversion and fabrication: Concentrated in Western Europe, the United States, Russia, and Japan.

- Transport and logistics: Under strict IAEA protocols, but vulnerable to sanctions, wars, and sabotage.

Russia, for instance, controls more than 40% of the global commercial uranium enrichment capacity and has agreements to build, supply, and even operate nuclear plants in dozens of countries — a geopolitical instrument disguised as energy cooperation.

Table 55: Strategic Services of the Nuclear Fuel Cycle

Service	Leading Countries	Remarks
Uranium Enrichment	Russia, France, USA, China	Russia controls ~40% of global capacity
Chemical Conversion	France, Canada, China	Pre-enrichment stage

Fuel Fabrication	Japan, France, Russia, USA	Production of nuclear fuel rods
Nuclear Transport	Various European countries, USA	Highly regulated by the IAEA

Source: Own elaboration based on the data presented in the Summary Table at the end of this chapter

The Risk of Dependence and the Race for Autonomy

The war in Ukraine dramatically exposed the risks of dependence on strategic uranium suppliers and nuclear services.

As a result, several countries:

- Relaunched domestic mining projects.

- Seek to diversify suppliers (e.g., Canada, Australia, Namibia).

- And aim to develop their own enrichment and fuel recycling capacities.

This new context has accelerated a race for nuclear energy security, particularly in Europe, the United States, and the Indo-Pacific region.

The Cycle as a Critical Chain: From Ore to Reactor

The nuclear supply chain is long, complex, and vulnerable. Therefore, securing each link is a national security priority.

A disruption at any stage (mining, enrichment, transport) can paralyze an entire national electrical system.

In the nuclear world, it is not enough to have technology. It is essential to have access to fuel — and independence in its transformation.

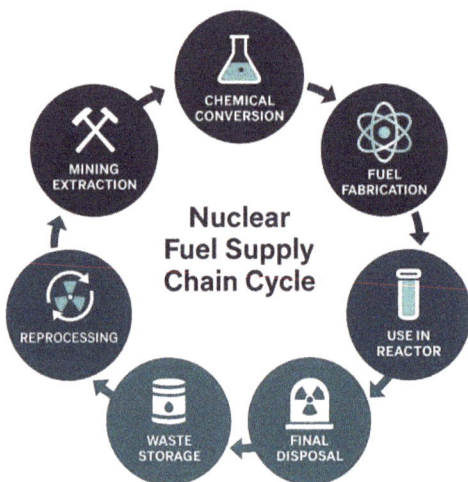

Illustrative infographic clearly showing the sequential cycle of the nuclear fuel supply chain.

Energy as a Weapon: The Case of Russia

The Russian Federation is not just a country with vast natural resources. It is an energy giant with its own geopolitical strategy, using energy as a tool of deterrence, influence, and political leverage.

Although natural gas has historically been Russia's primary energy weapon (with massive exports to Europe), the nuclear sector has emerged in recent decades as a new vector of silent but profound influence.

Rosatom: The Nuclear Arm of the Russian State

Rosatom, the Russian state corporation for nuclear energy, is much more than a company.

It acts as a geopolitical arm of the Kremlin, offering:

- Construction of complete nuclear reactors (turnkey).

- Favorable financing to partner countries.

- Continuous supply of nuclear fuel.

- And, in some cases, operation and maintenance of plants throughout their lifecycle.

With more than 70 active or planned international projects, Rosatom is the world's largest exporter of civil nuclear technology.

An Energy Influence Model

Russia's strategy is simple and effective:

1. Build nuclear power plants in developing countries (e.g., Turkey, Egypt, Bangladesh, Hungary).

2. Offer affordable financing and technical know-how.

3. Maintain exclusive fuel and technical service supply, creating long-term dependence.

This model is known as 'Nuclear Diplomacy' — a form of soft power with hard consequences.

From Cooperation to Dependence

Countries contracting Rosatom's plants become dependent on Russian-enriched uranium and logistics.

Even in European Union countries (like Hungary and Slovakia), Russian VVER-type reactors remain operational, with continuous supply from Moscow.

The energy crisis following the invasion of Ukraine (2022) demonstrated the weight of Europe's dependence on Russian energy — not just gas, but also nuclear.

Political Shielding through Nuclear Energy

Russia uses nuclear agreements to:

- Strengthen strategic alliances.

- Reduce the influence of Western powers in key regions (Middle East, Africa, Southeast Asia).

- Gain political support in international forums, exchanging technical assistance for votes or neutral positions in times of global tension.

In this context, nuclear energy is not just technology — it's diplomatic loyalty.

Risks and Tensions

Using nuclear energy as a geopolitical weapon carries risks:

- Instrumentalization of civil contracts for political purposes.

- Possibility of supply cuts in the event of sanctions or conflicts.

- Lack of quick alternatives for countries dependent on Rosatom.

This is why many countries are today:

- Rethinking their contracts with Russian suppliers.

- Attempting to diversify nuclear fuel sources.

- And developing national capacity to reduce vulnerability.

Russia has transformed nuclear energy into a powerful instrument of international influence.

Through Rosatom, it offers not only energy but a network of silent and lasting strategic dependence.

In a divided world, nuclear energy has become more than just science — it has become an extension of geopolitics.

Table 56: Global Presence of Rosatom and Russia's Nuclear Dependence

Country	Type of Cooperation	Remarks
Hungary	Construction and operation of VVER-1200 reactors (Paks II)	Russian financing and fuel supply
Turkey	Akkuyu plant (4 VVER-1200 reactors)	Turnkey project: Rosatom will operate for decades
Egypt	El-Dabaa plant (4 reactors)	Long-term agreement; partial Russian financing
Bangladesh	Rooppur plant (2 VVER-1200 reactors)	Technology, construction, and operation by Rosatom

India	Technical cooperation and construction (Kudankulam)	Russian reactors, in partnership with Indian companies
China	Various construction and technical cooperation agreements	Strategic relationship and joint R&D projects
Iran	Construction and operation of Bushehr	Technology transfer supervised by the IAEA
Vietnam	Planned (suspended)	The project was cancelled due to economic concerns
Algeria	Preliminary cooperation agreements	No project has started yet
Southeast Asia (various)	Negotiation of future projects	Strategic expansion ongoing

Source: Own elaboration based on the data presented in the Summary Table at the end of this chapter

The Role of the US, China, and France in the New Nuclear Race

Exporting nuclear power plants is much more than a technological transaction. It is a way to:

- Establish long-term relationships with foreign governments;

- Influence energy, industrial, and even diplomatic policies of partner countries;

- And compete for global spheres of influence, where energy, security, and scientific cooperation intertwine.

In the 21st century, this race has intensified. Russia dominates largely, but the US, China, and France are seeking to strengthen or recover their role in this silent and decisive chess game.

United States: Tradition, Decline, and Potential Rebirth

For decades, the US was a world leader in the export of nuclear technology. However:

- The private sector faced internal slowdown and external loss of competitiveness;

- High costs and regulatory delays hindered new projects;

- The merger of Westinghouse with foreign groups generated commercial instability.

In recent years, the US government has shifted strategically:

- Public support for the development of Small Modular Reactors (SMRs);

- Partnerships with strategic countries like Poland, Romania, and Ukraine;

- Active energy diplomacy to contain Rosatom's expansion.

Nuclear energy has returned to the US agenda as a piece of energy security and geopolitical influence.

China: Silent and Aggressive Rise

China is heavily investing in its domestic nuclear sector and seeks to export the Hualong One (HPR-1000) model to developing countries, offering:

- Generous state funding.
- Shorter delivery times.
- Associated infrastructure (training, equipment, logistical chains).

Ongoing or negotiating projects include:

- Pakistan (plants in operation and expansion).
- Argentina (technical negotiations).
- Several African and Southeast Asian countries.

China's strategy mirrors the model of the New Silk Road: energy in exchange for political alignment and economic openness.

France: Technical Tradition and Nuclear Diplomacy

France, through Orano (formerly Areva) and EDF, is historically a nuclear power with a strong technical reputation.

It exports European Pressurized Reactors (EPR), focusing on:

- The United Kingdom (Hinkley Point C),
- China (Taishan),
- India (Jaitapur, under negotiation).

The French approach is characterized by:

- High technical and environmental requirements.

- Less commercial aggressiveness compared to China or Russia.

- Nuclear diplomacy focused on Europe, Asia, and French-speaking Africa.

Table 57: Nuclear Technology Exporters and Geopolitical Strategies

Exporting Country	Main Technology	Export Strategy	Priority Target Countries
Russia	VVER (Rosatom)	Full financing, direct operation	Turkey, Egypt, Hungary, Bangladesh
China	Hualong One (HPR-1000)	Short deadlines, aggressive state support	Pakistan, Africa, Southeast Asia
USA	AP1000, SMRs	Strategic partnerships and security	Poland, Ukraine, Romania, Canada
France	EPR, EPR2	Technical tradition and European cooperation	United Kingdom, China, India, Francophone Africa

Source: Own elaboration based on the data presented in the Summary Table at the end of this chapter

The export of nuclear technology is today one of the most powerful tools for long-term international influence.

Each nuclear plant built represents decades of technical cooperation, supply contracts, workforce training, and political alignment.

In the 21st century's geopolitical chess game, nuclear energy is the battlehorse of the powers — and each reactor is a piece that conquers diplomatic territory.

Table 58: Nuclear Export Control Regimes

Regime	Objective	Key Members
Nuclear Suppliers Group (NSG)	Prevent proliferation via export control	48 countries (e.g., USA, UK, France)
Zangger Committee	List of controlled nuclear materials and technology	TNP Signatories
Wassenaar Arrangement	Control of sensitive goods and technologies	42 diverse countries

Source: Own elaboration based on the data presented in the Summary Table at the end of this chapter

Table 59: Nuclear Technology Export Flows by Country

Exporting Country	Main Destinations	Technology Systems
France	China, India	PWR, EPR reactors
Russia	India, Finland	VVER reactors
USA	UAE, Japan	AP1000 reactors

Source: Own elaboration based on the data presented in the Summary Table at the end of this chapter

Strategic Alliances and Energy Diplomacy

Contrary to what might seem, nuclear energy does not isolate — it brings countries closer. The technical complexity, risks, and costs involved require partnerships, mutual trust, and long-term political alignment. This is why nuclear energy is increasingly present in bilateral treaties, multilateral alliances, and free trade and defense forums.

The US–India Case: Nuclear Cooperation with Geopolitical Value

One of the most emblematic examples is the Civil Nuclear Cooperation Agreement between the United States and India, signed in 2008.

Pillars of the agreement:

- Recognition of India's rights to peaceful nuclear energy use, even though it is not a signatory of the NPT.

- Openness for civil nuclear technology trade and transfer.

- India's commitment to IAEA inspections at civil facilities.

Implicit objective:

- Strengthen the strategic partnership between the world's two largest democracies.

- Contain China's influence in Asia.

- Integrate India into the global civil nuclear system without requiring the dismantling of its military program.

This agreement set a historical precedent: a country outside the NPT but with good credentials became part of the 'global civil nuclear system.'

Energy and Free Trade Agreements

Nuclear energy is also starting to appear as a technical or strategic clause in multilateral agreements and regional treaties.

Examples:

- The USMCA (United States–Mexico–Canada Agreement) includes clauses on energy, infrastructure, and technological interconnectivity, with an indirect impact on nuclear supply chains.

- The EU–Japan and EU–South Korea agreements include technical components related to civil nuclear energy (standardization, safety, innovation).

These agreements often include:

- Secure technology transfer.

- Joint training/certification of operators.

- Integration into sensitive energy logistical chains.

Nuclear Energy and Security Alliances

Nuclear energy is also a tool for consolidating military alliances, even when the reactors are civilian.

NATO:

- Many NATO countries use shared nuclear technology with the US.

- There are nuclear sharing protocols that include weapons but also strategic civil cooperation (Germany, Netherlands, Belgium, Italy).

AUKUS (Australia–UK–US):

- A defense agreement that includes the transfer of nuclear submarine technology to Australia.

- Marks the entry of military nuclear technology into 21st-century alliances, even though Australia remains without nuclear weapons.

Quad (US–India–Japan–Australia):

Cooperation in Indo-Pacific security involving energy partnerships, including civil nuclear energy.

Energy as the Axis of 21st-Century Technological Alliances

In the 21st century, energy security and technological innovation are the new pillars of global diplomacy.

And civil nuclear energy is at the center of this movement.

Table 60: Strategic Agreements and Nuclear Diplomacy

Agreement / Alliance	Participants	Type of Cooperation	Strategic Objective
US–India Agreement (2008)	USA and India	Civil nuclear cooperation outside the NPT	Bring India closer to the West and contain China
NATO Nuclear Sharing	USA and European NATO countries	Weapons sharing and technical cooperation	Deter nuclear threats, integrate allies

AUKUS	Australia, United Kingdom, USA	Nuclear submarines and military cooperation	Counterbalance Chinese influence in the Indo-Pacific
Quad (Security Dialogue)	USA, Japan, India, Australia	Energy and technology partnership	Regional stabilization and security
EU–Japan / EU–South Korea	European Union and Asian partners	Civil nuclear technology and standards	Technical and energy integration
USMCA (former NAFTA)	USA, Mexico, Canada	Energy integration and infrastructure	Energy security and supply chain logistics

Source: Own elaboration based on the data presented in the Summary Table at the end of this chapter

Table 61: Impact of Civil Nuclear Energy

Theme	Impact of Civil Nuclear Energy
Energy Transition	Carbon-free generation and foundation for renewables
Scientific Cooperation	Training of personnel, joint R&D, innovation
Technological Security	Sensitive and secure supply chains
Strategic Autonomy	Reduced dependence on fossil fuels
Bilateral Diplomacy	Tool for rapprochement and mutual trust

Source: Own elaboration based on the data presented in the Summary Table at the end of this chapter

In the 20th century, nuclear energy was a weapon. In the 21st century, it is diplomacy, innovation, and sovereignty.

Nuclear energy has become a vector of strategic rapprochement between nations, integrating treaties, military alliances, and technological agreements.

Each nuclear contract is more than energy — it is a bridge between countries.

Table 62: Geopolitical Impact of Nuclear Projects

Project	Host Country	Key Impacts
Russia–India Nuclear Deal	India	Strengthened alliances and technology dependence
EPR in China	China	Technological advancement and EDF partnerships
Barakah (UAE)	UAE	Energy diversification and energy diplomacy

Source: Own elaboration based on the data presented in the Summary Table at the end of this chapter

The New Global Energy Order and the Role of Nuclear Energy

The 21st century has brought a profound transformation in global energy paradigms:

- Decarbonization has become a climatic and political imperative.

- Energy autonomy has become a matter of national security.

- Competition for clean and strategic technologies has intensified among major powers.

In this context, nuclear energy reemerges as a central — and often underestimated — pillar of this new global energy order.

Nuclear Energy Between Climate Emergency and Realpolitik

For decades, nuclear energy was debated in moral, technical, or emotional terms. Today, it is evaluated in light of climatic urgency and strategic security.

From a climatic point of view:

- It is one of the only sources capable of generating stable, large-scale, emission-free electricity.

- It complements intermittent renewables (solar, wind) with a firm baseload.

From a geopolitical point of view:

- It reduces dependence on imported fossil fuels.

- It lowers vulnerability to external energy shocks.

- It strengthens the technological and industrial sovereignty of States.

Result: countries that once hesitated are now reconsidering or relaunching their nuclear programs.

Table 63: Evolution of Nuclear Policies by Country or Bloc

Trend	Countries / Blocs
Total nuclear phase-out	Germany, Belgium
Strategic reconsideration	Japan, Italy, Spain
Maintenance or expansion	France, United Kingdom, USA
Aggressive expansion	China, Russia, India, South Korea
New entrants	Brazil, Turkey, Egypt, Bangladesh

Source: Own elaboration based on the data presented in the Summary Table at the end of this chapter

The global energy transition is not homogeneous — it is multipolar, asymmetrical, and deeply geopolitical.

The Role of Nuclear Energy in the New Energy Matrix

The energy matrices of the near future will be based on four complementary pillars:

1. Intermittent renewables (solar, wind, hydro).

2. Dispatchable low-emission sources (such as nuclear).

3. Smart and interconnected electric grids.

4. Energy storage and green hydrogen.

In this model, nuclear energy fulfills two essential functions:

- Stabilizer of the electric system.

- Strategic reserve in times of crisis or scarcity.

Chart 50: Global Energy Investment: Nuclear vs Renewables (2015–2023)

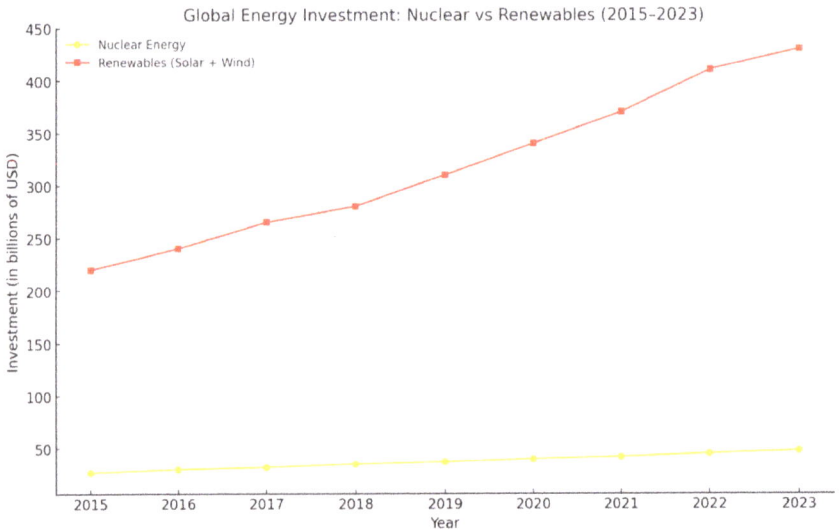

Source: Own elaboration based on the data presented in the Summary Table at the end of this chapter

A Factor of Geopolitical and Industrial Influence

- In a world where:

- • Technology is sovereignty,

- • Resources are weapons,

- • And energy is diplomacy,

Nuclear energy is a multiplier of national power.

Whether as an energy source, an export vector, or a diplomatic tool, the country that masters nuclear technology also masters part of the future.

Nuclear energy is no longer just a topic for engineers or activists. It is a key piece in the dispute for the 21st century.

At the heart of the new global energy order, nuclear energy represents not only electricity — but autonomy, security, and influence.

Table 64: Geopolitical Blocs and Their Stance on Nuclear Energy

Bloc / Region	Current Stance	Remarks
European Union	Mixed / Divided	France leads expansion; Germany and others step back.
North America	Strategic reinforcement	The USA invests in SMRs; Canada invests in international partnerships.
Asia-Pacific	Strong expansion	China, India, and South Korea are experiencing ongoing growth.
Latin America	Moderate growth	Brazil strengthens the sector with naval and civil projects.
Africa	Beginning integration	Egypt leads with Russian support and growing interest in several countries.

Middle East	Energy diversification	UAE, Iran, Saudi Arabia, and Turkey invest in nuclear.

Source: Own elaboration based on the data presented in the Summary Table at the end of this chapter

Conclusion of the Chapter – Nuclear Geopolitics: Energy, Power, and Global Influence

Nuclear energy, more than just a source of electricity, is today a symbol of sovereignty, prestige, and strategic autonomy. Since its origin, it has served as a foundation of international power — sometimes as an instrument of deterrence, sometimes as a tool of technical cooperation, or as a vector of diplomatic influence.

In this chapter, we saw that:

- Mastering the nuclear cycle can confer geopolitical respect even without nuclear weapons (latent deterrence).

- The UN Security Council crystallizes the monopoly of those who first mastered the atom.

- Treaties like the NPT and organizations like the IAEA shape access to technology — not always impartially.

- Control of uranium and supply chains has become a new strategic front.

- Russia, China, the US, and France today compete for global influence through the export of nuclear technology.

- strategic alliances like AUKUS, NATO, and the US–India agreements show that nuclear energy is also diplomacy and security.

In a multipolar and transitional world, those who control nuclear energy not only light up cities — they influence nations.

This chapter demonstrates that nuclear energy no longer belongs solely to the technical domain but is part of the great game of power that defines the 21st century.

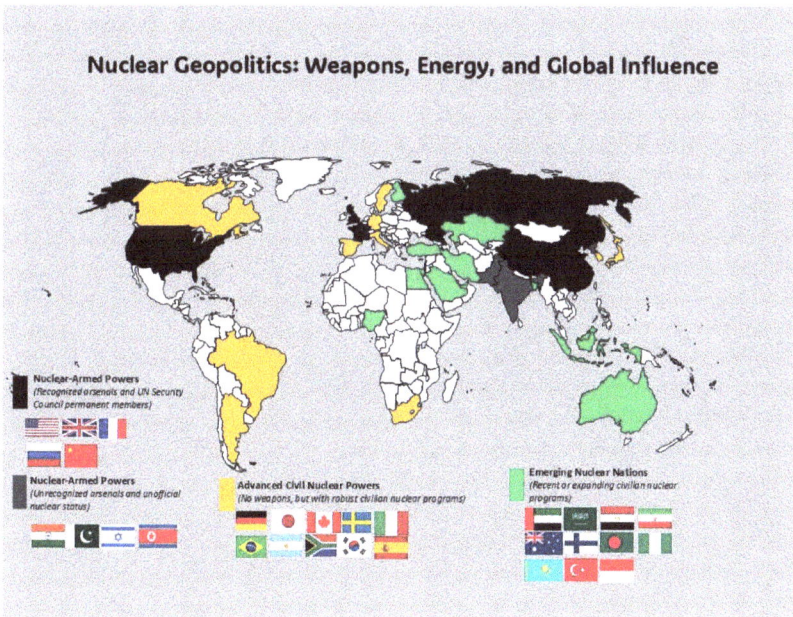

Nuclear Geopolitics: Weapons, Energy, and Global Influence

Table 65: Sources Consulted in Chapter 7

Source	Reference
International Atomic Energy Agency (IAEA)	Reports and official publications on nuclear energy, non-proliferation, and safeguards.
Nuclear Energy Agency (NEA/OECD)	Studies on nuclear policy, technology, and geopolitical implications.
World Nuclear Association (WNA)	Data on uranium production, nuclear reactors, and international supply chains.
US Department of Energy (DOE)	Reports on US nuclear energy policy, SMRs, and technological innovation.
International Energy Agency (IEA)	Global energy outlooks, including the role of nuclear energy in the energy transition.
Rosatom State Atomic Energy Corporation	Public materials on international projects and Russia's nuclear export strategy.
World Nuclear News (WNN)	News and updates on global nuclear developments and industry trends.
French Alternative Energies and Atomic Energy Commission (CEA)	Publications on nuclear technology, innovation, and diplomacy.
Energy Information Administration (EIA)	Statistics and reports on nuclear generation, uranium resources, and forecasts.

Ministry of Energy of the People's Republic of China	Strategic documents on the expansion of nuclear energy in China.
European Commission	Documents on the European Union's position on nuclear energy and supply security.
Academic Articles and Specialized Journals	Various scientific articles analyze nuclear geopolitics, energy security, and technological diplomacy.

Preparation for the Next Chapter – Nuclear Energy in the World: Who Won and Who Lost?

A Strategic Comparison of Nuclear Choices

Over time, each country has adopted different strategies regarding nuclear energy:

- Some invested heavily and are now reaping the benefits in the form of clean energy, energy security, and technological leadership.

- Others succumbed to fear or political pressure, dismantled their programs — and now face high costs, external dependency, and industrial fragility.

This new chapter proposes a comparative and objective analysis accompanied by emblematic examples.

It seeks to answer whether energy choices were sovereign decisions... or strategic mistakes. We will dive into several cases to understand the real impact of national decisions — both in the present and in the future.

Chapter 8 –
Nuclear Energy Around the World:
Who Won and Who Lost?

This chapter will compare four distinct strategic energy models adopted by countries that:

- Abandoned nuclear energy for political or ideological reasons.

- Cautiously invested based on technical and scientific criteria.

- Heavily invested with clear goals of sovereignty and development.

- Refused to even consider nuclear as an alternative.

The reader will be invited to reflect on the practical, economic, climatic, and geopolitical consequences of these choices.

In the vast energy landscape of the 21st century, each country builds its path based on national priorities, perceptions of risk, political pressures, and strategic ambitions. And among all the choices a state can make, few are as decisive and symbolic as the option — or rejection — of nuclear energy.

While some nations see nuclear as a pillar of energy security, sovereignty, and climate transition, others view it as a latent threat, a political burden, or an unwanted legacy of the Cold War. The decisions made over the past decades reveal deep divergences between countries, and their effects have now become visible, measurable — and, in some cases, irreversible.

In this chapter, we propose a comparative reflection between four countries that followed radically different energy strategies:

• **Germany:** A country that, after the Fukushima accident, decided to completely abandon nuclear energy, betting on renewables and natural gas — with significant economic and climate consequences.

• **Finland:** A nation that, with technical prudence and political consensus, maintained and expanded its nuclear program, achieving energy security and low emissions in a stable and transparent model.

• **United Arab Emirates:** A surprising case of a petroleum-exporting country that, with vision and speed, decisively invested in nuclear as a symbol of modernity, sustainability, and international projection.

• **Portugal:** A European example of total rejection of nuclear energy, focusing exclusively on renewables — an idealistic and environmentally well-intentioned choice, but one that raises questions about autonomy, intermittency, and technological balance.

These four trajectories reveal not only technical decisions but distinct visions of the future. Behind each energy model are factors such as:

- Political and environmental culture,
- Economic structure,

- Technical and scientific capacity,

- Geography and available natural resources,

- And above all, the perception of long-term risk vs. benefit.

The question posed by this chapter is simple yet crucial:

Who won and who lost?

Did those who ensured energy stability and low emissions win?

Did those who preserved political independence from major suppliers win?

Or did those who gained public opinion, even at high economic costs, win?

Throughout the following sections, we will analyze real data, public policies, and tangible outcomes. Not to judge but to understand. Because, in the end, energy choices are always a reflection of a nation's priorities and values.

Germany – The Abandonment of Nuclear Energy and the Rise of Vulnerability

A historic shift with global implications

In 2011, following the Fukushima accident, Germany announced one of the most radical energy decisions of the 21st century: the total shutdown of its nuclear power plants by 2023.

Motivated by a strong anti-nuclear sentiment already present in German society since Chernobyl, Chancellor Angela Merkel — a scientist by training and until then a supporter of nuclear energy — completely reversed the country's energy policy, launching the so-called *Energiewende* (Energy Transition).

The bet to replace nuclear energy with renewable sources (wind and solar) supported temporarily by natural gas.

The goal is to decarbonize the economy without compromising energy security.

The result: much more ambiguous than expected.

Immediate consequences and structural contradictions

Despite the environmental rhetoric, the abandonment of nuclear power generated contradictory practical effects:

- Increase in CO_2 emissions: The reduction in nuclear capacity was partially offset by greater use of coal and lignite, both highly polluting.

- Dependence on Russian gas: To maintain grid stability, Germany reinforced contracts with Gazprom and became highly dependent on imported energy from Russia.

- Grid instability: The intermittency of renewables required massive subsidies, backup mechanisms, and electricity imports — mostly from France (which used... nuclear energy).

- High costs for consumers: Energy prices rose significantly, putting pressure on both industry and households.

- Growing industrial fragility: Energy-intensive companies began relocating production to countries with lower costs, including... China.

The country that vowed to lead the energy transition became dependent on Russian fossil fuels and French nuclear energy.

Chart 51: Evolution of CO_2 Emissions in Germany

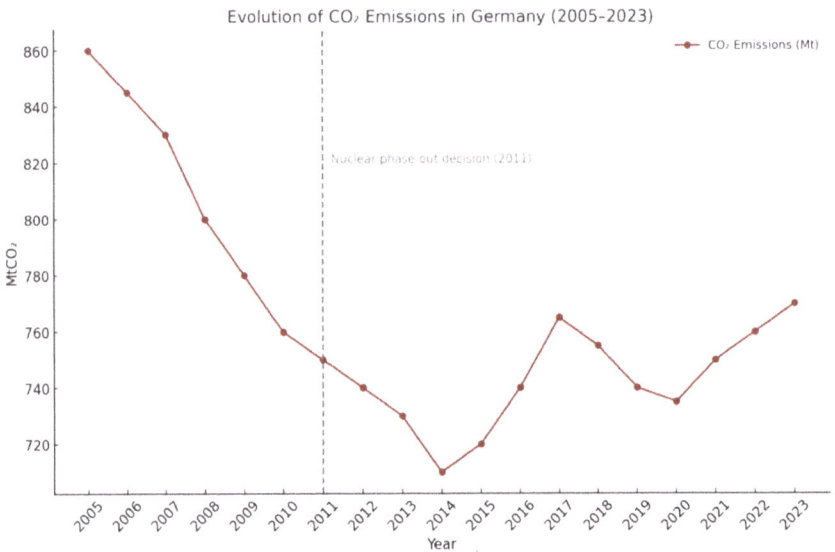

Evolution of CO_2 Emissions in Germany (2005-2023)

Source: Own elaboration based on the data presented in the Summary Table at the end of this chapter

Graph of CO_2 Emissions Evolution in Germany (2005–2023), highlighting the year of the nuclear phase-out decision (2011).

Chart 52: Evolution of the Electricity Mix in Germany

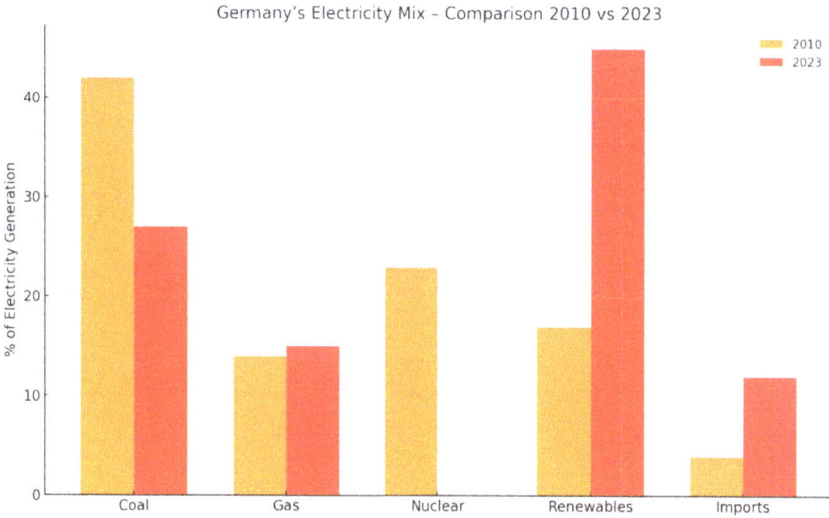

Germany's Electricity Mix – Comparison 2010 vs 2023

Source: Own elaboration based on the data presented in the Summary Table at the end of this chapter

Comparative graph of Germany's electricity mix between 2010 (pre-phase-out) and 2023 (post-phase-out).

Energy Geopolitics Exposed by War

The invasion of Ukraine in 2022 marked a turning point. Germany found itself with no room to maneuver, facing:

- The urgent need to reduce Russian imports.

- The inability to compensate with renewables in time.

- The absence of nuclear power plants that could offer domestic electricity security.

The decision to abandon nuclear energy proved to be, from a geopolitical standpoint, a serious strategic vulnerability.

Chart 53: Evolution of Germany's External Energy Dependence

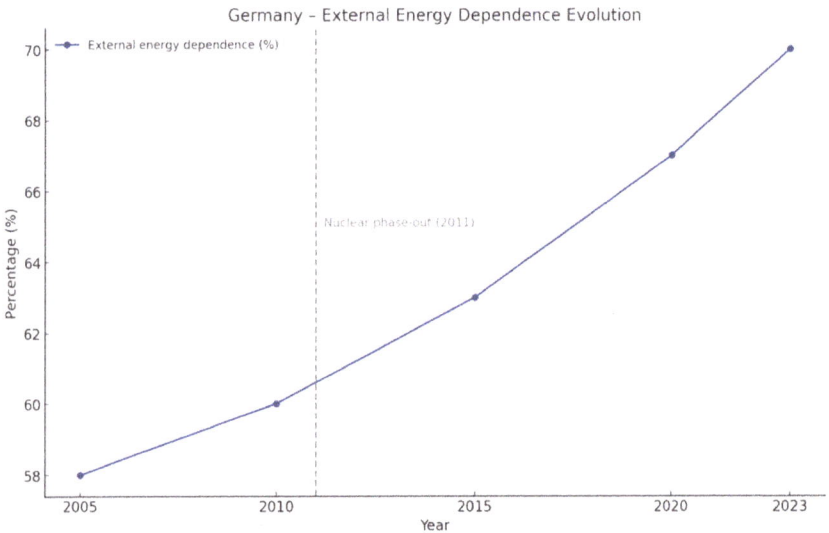

Source: Own elaboration based on the data presented in the Summary Table at the end of this chapter

Graph showing the evolution of Germany's external energy dependence, highlighting the post-nuclear phase-out period.

A Debate Reignited... Too Late?

Today, various experts and political sectors in Germany are questioning whether the complete shutdown of nuclear power plants was:

- An emotional and hasty response to post-Fukushima fears.

- Or a missed historic opportunity to lead Europe's energy transition with balance.

Even with increasing pressure to reverse the decision, the dismantling of infrastructure, the loss of technical human capital, and years of disinvestment make any return highly unlikely.

Table 66: Summary of Impacts

Indicator	Current Situation (2023–2024)
CO_2 Emissions	Higher than in 2010
Electricity Mix	High intermittency, with residual coal use
Energy Imports	Very high, especially gas and electricity
Consumer Costs	Among the highest in Europe
Public Satisfaction	Divided; growing skepticism.
Industry	Under pressure, with relocations abroad

Source: Own elaboration based on the data presented in the Summary Table at the end of this chapter

Germany is a striking example of how political decisions driven by ideology and public pressure — even if well-intentioned — can lead to deeply counterproductive practical consequences.

Chart 54: Evolution of Electricity Prices in Germany

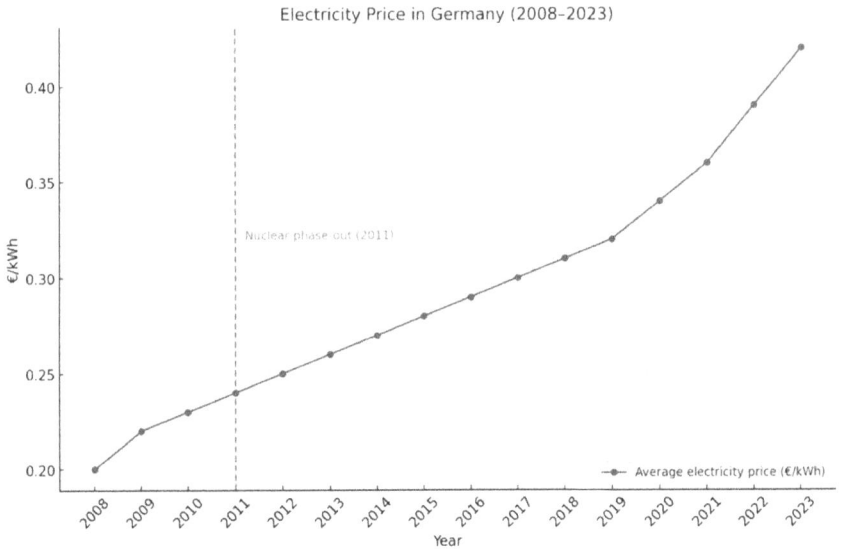

Electricity Price in Germany (2008–2023)

Source: Own elaboration based on the data presented in the Summary Table at the end of this chapter

Graph of electricity price evolution in Germany (2008–2023), highlighting the post-nuclear phase-out period.

The nuclear phase-out, far from representing an ecological breakthrough, led to dependence, instability, and inconsistency — in a country that had pledged to lead Europe's energy future.

Finland – Technical Persistence and Strategic Security

A consistent commitment to science, safety, and stability.

Summary of Indicators:

Table 67: Key Indicators of Finland's Energy Policy

Indicator	Current Situation (2023–2024)
CO_2 Emissions	Among the lowest per capita in the EU
Electricity Mix	>40% nuclear, >40% hydroelectric
Energy Imports	Minimal; high autonomy
Electricity Costs	Stable, moderately high (Nordic standard)
Public Satisfaction	High; growing support for nuclear
Waste Management	World reference (Onkalo)

Source: Own elaboration based on the data presented in the Summary Table at the end of this chapter

A Model of Pragmatism and Long-Term Vision

While some countries retreated in the face of ideological pressures or media-driven disasters, Finland adopted a different approach: pragmatic, scientific, and long-term.

In a global energy landscape marked by uncertainty, Finland chose the path of technical resilience, maintaining and modernizing its nuclear program with broad political and social support.

Today, Finland is considered a model of success in civilian nuclear management, combining:

- Rigorous operational safety.

- Institutional transparency.

- And public trust sustained by decades of consistency.

Chart 55: Evolution of Finland's Energy Mix

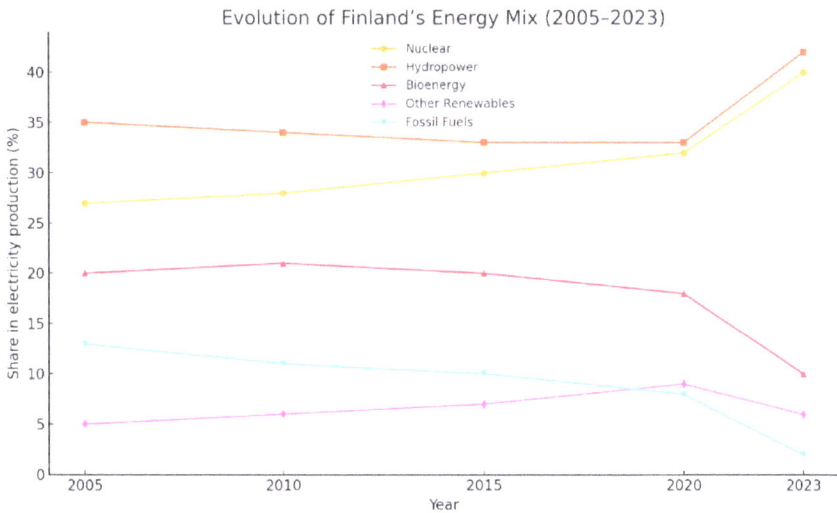

Evolution of Finland's Energy Mix (2005-2023)

Source: Own elaboration based on the data presented in the Summary Table at the end of this chapter

Graph showing the evolution of Finland's energy mix (2005–2023), clearly illustrating the growth of nuclear share and the decline of fossil fuels.

The Finnish Nuclear Program: Pillars of a Stable Policy

Finland began its nuclear journey in the 1970s. Today, it operates five nuclear reactors, the most recent of which — Olkiluoto 3 — came online in 2023, becoming the largest operating reactor in Europe.

Key aspects of Finland's approach:

- Partial technical autonomy: Although dependent on consortia and imports, Finland has strong national agencies and an advanced technical culture.

379

- Strict waste management: It is a global pioneer in deep geological storage (Onkalo project), serving as a benchmark in safety and innovation.

- Public consultation and social acceptance: The decision-making process included public hearings, independent studies, and information campaigns, promoting transparency and consensus.

In Finland, nuclear energy is viewed as part of the solution — not as a problem to avoid.

Tangible benefits of the nuclear policy

Finland's energy policy has generated measurable gains:

- Low carbon emissions: Around 90% of Finland's electricity is emission-free, combining nuclear and hydropower.

- High energy security: Finland produces almost all of its electricity domestically, reducing external vulnerabilities.

- Tariff stability: Although energy prices are high due to Nordic factors, they remain stable and predictable.

- Growing confidence in nuclear energy: According to recent polls, over 60% of the population supports nuclear power — one of the highest levels in Europe.

Chart 56: Evolution of CO$_2$ Emissions in Finland

CO$_2$ Emissions in Finland (2005-2023)

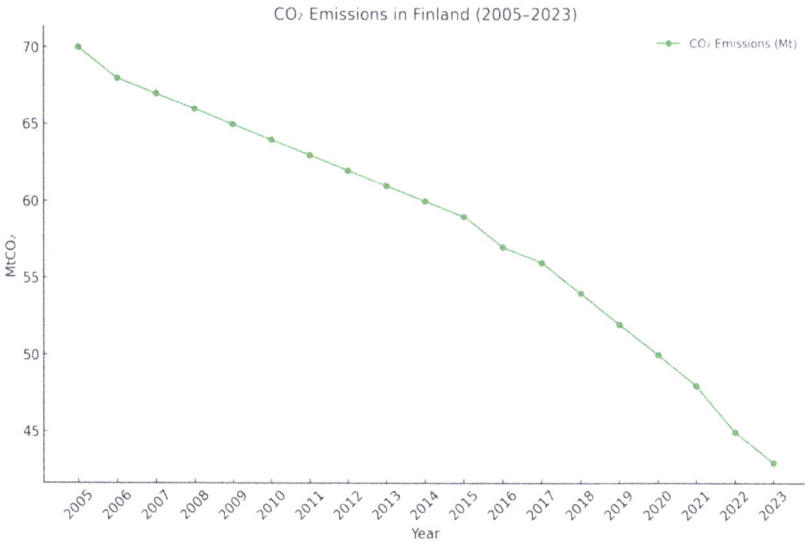

Source: Own elaboration based on the data presented in the Summary Table at the end of this chapter

Graph showing the evolution of CO$_2$ emissions in Finland (2005–2023) — with a clear and consistent downward trend over the years.

Chart 57: Evolution of Finland's Energy Dependence

Evolution of Finland's Energy Dependence

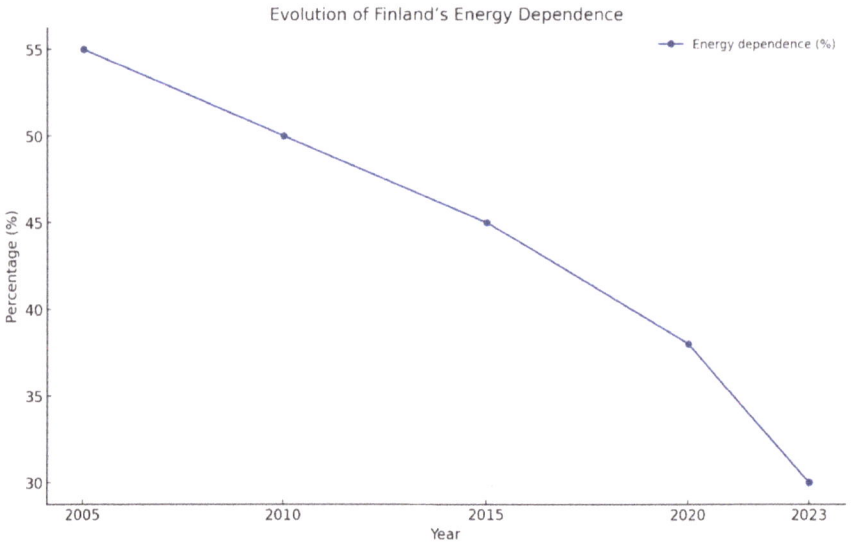

Source: Own elaboration based on the data presented in the Summary Table at the end of this chapter

Graph showing the evolution of Finland's energy dependence, highlighting a clear reduction over the past two decades.

Explanation – Why is Electricity More Expensive in the Nordic Countries?

Higher electricity prices in countries like Finland, Denmark, and Sweden are not directly related to energy inefficiency or external dependence — quite the opposite. These prices result from a set of structural and cultural factors, informally known as the **"Nordic factor"**:

- High environmental and energy taxes – levied to fund infrastructure and encourage efficiency.

- Robust and decentralized electrical grid – costly to maintain due to long distances and low population densities.

- High standards of reliability and quality – requiring continuous investment.

- Extreme climate – which demands heavy use of electric heating and constant backup.

- Export to European markets – local prices are partly influenced by European market mechanisms and regional exchanges.

In short, one pays more for cleaner, more reliable electricity with a lower environmental footprint — the price of Nordic technical and climate excellence.

Average Residential Electricity Prices in the EU (2023)

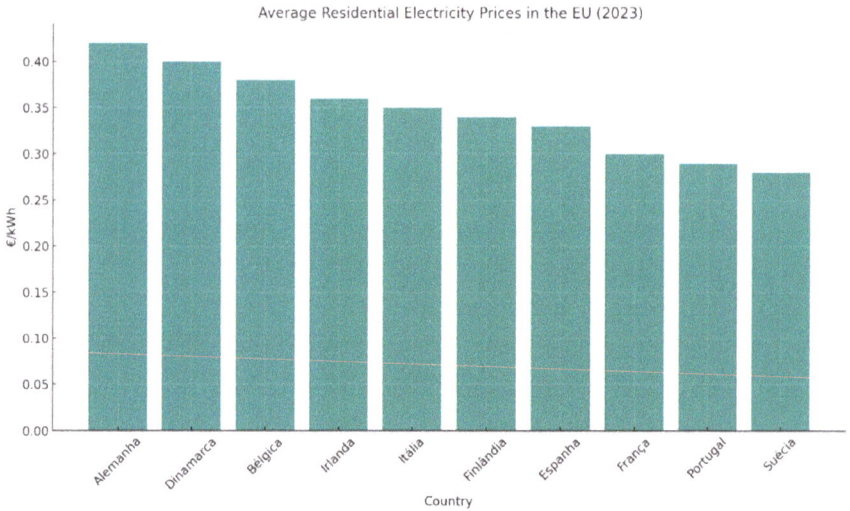

Source: Own elaboration based on the data presented in the Summary Table at the end of this chapter

Comparative graph of household electricity prices in the European Union (2023), showing Finland with high values, yet still below the peaks seen in countries like Germany and Denmark.

The Olkiluoto 3 Project: Technical Challenge, Political Victory

The Olkiluoto 3 (EPR) reactor is an ambitious project:

- It faced significant delays and cost overruns, which initially led to criticism.

- But once it finally came online, it became a European engineering landmark.

- It strengthened Finland's energy base and further reduced the need for imports.

Olkiluoto 3 Reactor Output:

Olkiluoto 3 is a European Pressurized Reactor (EPR) with a net output of:

- 1,600 MW (megawatts)
- It is currently the largest operational nuclear reactor in Europe

This reactor alone can produce approximately 14% of all electricity consumed in Finland — a clear demonstration of nuclear power's scale and stability in a country with a harsh climate.

This case shows that, even with challenges, technical perseverance pays off — especially when supported by strategic vision and public backing.

Finland demonstrates that nuclear energy, when well-planned and managed, is not a risk — it is a strategic advantage.

By resisting ideological pressure and focusing on science, it has:

- Reduced emissions,
- Ensured stability,
- And prepared itself for the future with resilience.

While others retreated in fear, Finland advanced methodically — and is now reaping the rewards of that persistence.

United Arab Emirates – Accelerated Commitment, Visible Results

Bringing in the visionary spirit of Sheikh Mohammed bin Rashid Al Maktoum — especially through the metaphor of the "lion and the gazelle" — adds strength, identity, and emotion to this exemplary case of strategic development.

"Every morning in Africa, a gazelle wakes up. It knows it must run faster than the fastest lion, or it will be killed.

Every morning, a lion wakes up. It knows it must outrun the slowest gazelle, or it will starve.

It doesn't matter whether you are the lion or the gazelle — when the sun comes up, you'd better be running."

— My Vision, Sheikh Mohammed bin Rashid Al Maktoum

This vision of constant movement, anticipation, and courage perfectly fits what the United Arab Emirates (UAE) embodied when they decided to invest in nuclear energy: a small country but agile, ambitious, and determined not to rely on chance — but on their own vision.

The Gazelle and the Lion: The Race for Progress

"It doesn't matter whether you are the lion or the gazelle — when the sun comes up, you'd better be running."

— Sheikh Mohammed bin Rashid Al Maktoum, My Vision

With this powerful metaphor, Dubai's visionary leader describes the nature of progress: there is no room for inertia in the 21st century. In a world of global competition, countries that wish to secure prosperity for future generations must run — strategically, boldly, and with vision.

And that is precisely what the United Arab Emirates has done over the past two decades. From a country traditionally reliant on oil exports, a nation emerged that bets on innovation, sustainability, and energy independence.

The choice for nuclear energy was rapid, well-executed, and strategically crafted — an example of how political will, coupled with efficient technical management, can transform realities.

The Barakah Project: Vision, Speed, and Execution

In 2008, the UAE announced the creation of a civil nuclear program with very clear goals:

- Diversify the energy matrix.
- Preserve fossil resources for export.
- Reduce carbon emissions.
- Train a new generation of national engineers and technicians.

Shortly after, in 2009, the KEPCO consortium (South Korea) won the bid to construct four APR-1400 reactors in Barakah, in the desert region of Al Dhafra.

Results:

- 2020 to 2023: The four reactors were commissioned sequentially, within adjusted timelines.
- Total installed capacity: around 5,600 MW, equivalent to 25% of the country's electricity.

- Zero local CO_2 emissions during operation.

- Establishment of an independent regulatory agency (FANR) and training of hundreds of national professionals.

Integrated Energy Strategy

Unlike many countries that treat nuclear energy as an isolated alternative, the UAE incorporated it as part of a comprehensive energy plan.

The Energy Strategy 2050 foresees:

- Increasing renewables (solar) to complement nuclear.

- Rational use of natural gas.

- Promotion of energy efficiency in buildings, transport, and industry.

- Training national professionals and forging partnerships with international universities.

The Emirates did not view nuclear energy as an end but as a means to achieve stability, innovation, and leadership.

An Exemplary Case in Arab Geopolitics

Chart 59: Evolution of the Electricity Mix in the UAE

Evolution of the Electricity Mix in the United Arab Emirates (2010-2023)

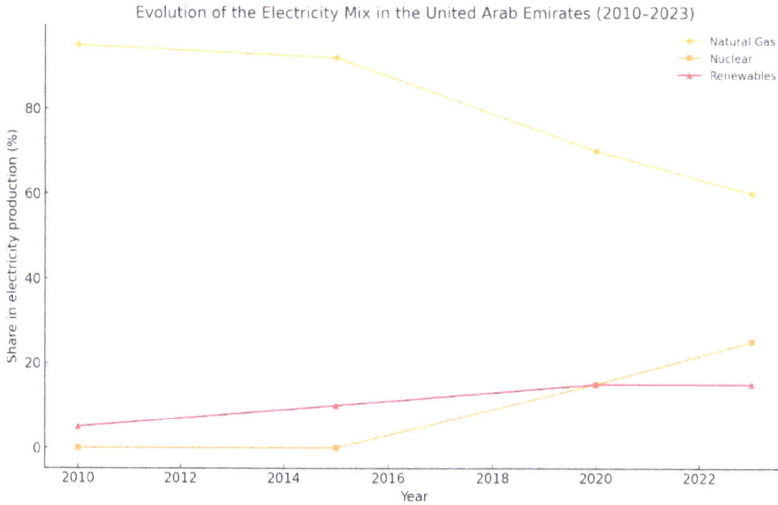

Source: Own elaboration based on the data presented in the Summary Table at the end of this chapter

Graph showing the evolution of the electricity mix in the United Arab Emirates (2010–2023), highlighting the growing share of nuclear energy and the decline of natural gas.

The UAE's entry into the club of civil nuclear nations did not go unnoticed:

- They became the first Arab country to operate large-scale commercial reactors.

- They established international trust through a clear commitment to non-proliferation (NPT, IAEA).

- They strengthened their regional geopolitical influence by demonstrating that technological and energy

389

development can go hand in hand with political stability.

Key Indicators

Table 68: Indicators of the UAE's Energy Policy

Indicator	Current Situation (2023–2024)
CO_2 Emissions	Significant reduction thanks to nuclear
Electricity Mix	25% nuclear, 10% renewables, remaining natural gas
Energy Imports	Virtually none; electrical self-sufficiency
Electricity Costs	Stable and regionally competitive
Public Satisfaction	High; nuclear well received in society
Technology Transfer	High cooperation with South Korea and the USA

Source: Own elaboration based on the data presented in the Summary Table at the end of this chapter

Leadership, Vision, and Execution: The UAE Example

The case of the United Arab Emirates shows that when there is leadership, vision, and execution, the impossible becomes possible.

In a scenario where many still hesitate, the UAE chose to run — like the gazelle and the lion of Sheikh Mohammed — and, in that race, achieved what many older or wealthier nations have yet to accomplish: a modern, clean, and sovereign energy matrix.

The success of the nuclear program in Barakah is not just an energy achievement — it is the expression of a nation that chose to take control of its destiny.

Chart 60: Evolution of CO_2 Emissions in the UAE

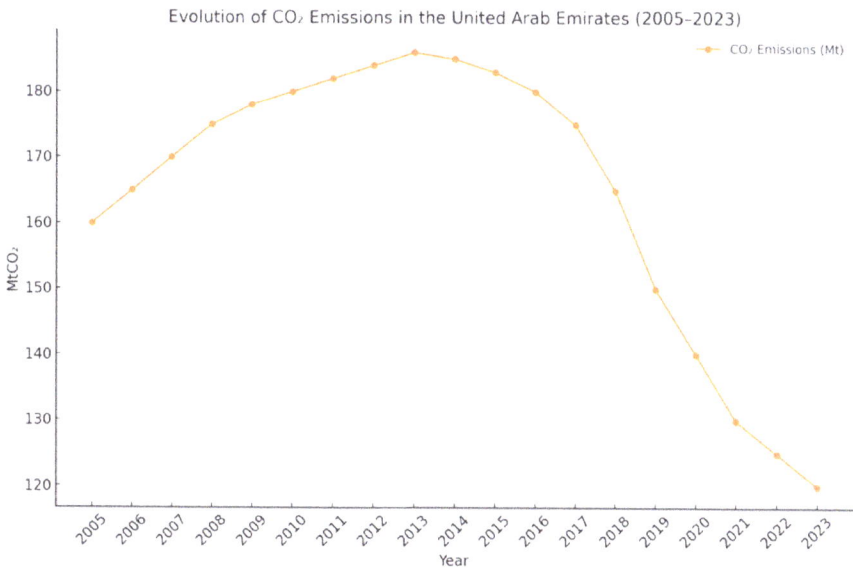

Evolution of CO_2 Emissions in the United Arab Emirates (2005-2023)

Source: Own elaboration based on the data presented in the Summary Table at the end of this chapter

Graph showing the evolution of CO_2 emissions in the United Arab Emirates (2005–2023) — highlighting a clear reduction following the introduction of the nuclear program.

Why is electricity more expensive in the UAE than in neighboring countries?

Chart 61: Average Household Electricity Prices in the Gulf Region

Average Residential Electricity Prices in the Gulf Region (2023)

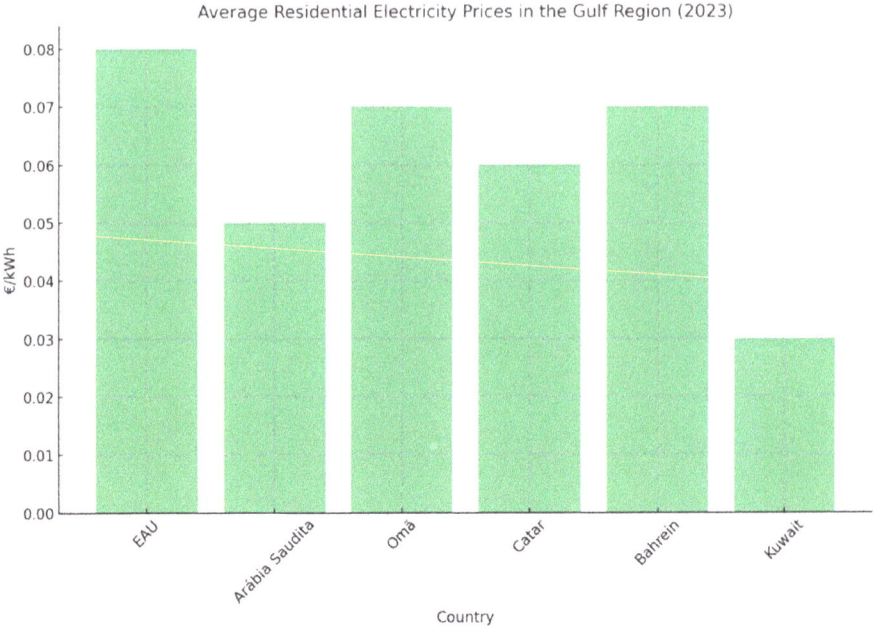

Source: Own elaboration based on the data presented in the Summary Table at the end of this chapter

Comparative graph of household electricity prices in the Gulf region (2023) — highlighting the United Arab Emirates as one of the countries with stable and competitive rates within the regional context.

At first glance, it might seem contradictory that the United Arab Emirates has one of the highest electricity tariffs in the Gulf while at the same time:

- Attracting major industrial and agricultural investments.

392

- Maintaining high regional competitiveness.

- And remaining a benchmark for energy stability and modern infrastructure.

The explanation requires technical, economic, and strategic context — and can be structured on three levels:

1. A More Realistic and Rational Pricing Policy

While many Gulf countries still maintain widespread and distortionary electricity subsidies (such as Kuwait and Saudi Arabia), the UAE has, in recent years, adopted a policy of gradual tariff reforms, aligning prices more closely with the real cost of production — a more sustainable approach in the long term.

In the UAE, consumers pay more... but the energy structure is more transparent, efficient, and resilient.

2. High Standards of Quality, Stability, and Coverage

The price also reflects:

- The enormous installed reserve capacity (to guarantee stability under desert climate conditions).

- The rapid modernization of the electric grid, digitalization, and constant maintenance.

- Investment in infrastructure resilient to extreme heat, sand, and corrosion.

In addition, the country has a more detailed, consumption-tiered tariff system — giving the illusion of high prices, even though basic consumption remains accessible.

3. Strategic Vision: Energy as a Pillar of Development

Despite slightly higher costs compared to its neighbors, the UAE offers:

- Total energy stability (no blackouts).

- Tariff predictability (no surprises or crises).

- Integration of clean sources (nuclear and solar).

- A clear regulatory environment that attracts investor confidence.

In other words, what investors gain is not just cheap energy but reliable, clean, and accessible energy, along with institutional and geopolitical stability.

For those seeking to grow tomatoes in the desert or set up a high-tech factory, this is worth much more than a few cents per kWh.

Comparison with the Global Scenario

Even though electricity is more expensive regionally, prices in the UAE are far lower than in Europe or Asia. And, if we exclude the hidden subsidies of neighboring countries, the UAE has a much more rational and sustainable energy structure.

Although electricity prices in the UAE are among the highest in the Gulf, they should be seen as a sign of energy maturity and economic attractiveness. The country has created an environment where investors are assured of the following:

- 24-hour energy availability,

- Excellence in infrastructure,

- Energy diversification (nuclear + solar),

- And political and economic stability.

The UAE does not subsidize energy — they subsidize the future.

Analytical Complement: Electricity Prices in the UAE – Cost or Strategy?

A recurring observation is that the UAE has higher electricity tariffs than its Gulf neighbors, such as Saudi Arabia or Kuwait. However, this cost difference should not be interpreted as an economic weakness but rather as part of a more mature energy and institutional strategy.

While several countries in the region still operate under widespread and unsustainable subsidies, the UAE has opted for a more transparent, tiered, and technically sound approach:

- Prices reflect the real cost of generation and distribution — including nuclear, solar energy, and world-class infrastructure.

- The national grid is resilient to extreme weather, highly digitalized, and extensively modernized.

- The country promotes efficiency and tariff predictability to ensure long-term stability, even with growing demand.

Moreover, the tariff differential is offset by a reliable institutional environment, legal security, logistical connectivity, and an internationally consolidated reputation as a hub for business, technology, and innovation.

The Emirates offer not just energy — they offer reliability, predictability, and strategic vision.

This explains why the country attracts agricultural investments in desert zones, high-energy industrial parks, data centers, and high-tech laboratories — even with a cost per kWh higher than neighboring countries.

In the global context, UAE electricity prices remain well below those in Europe or Asia, reinforcing the country's attractiveness.

Table 69: Household Electricity Prices Across Various Countries

Country	Average Price (€/kWh) in 2023
Germany	0.42
Portugal	0.29
Finland	0.34
UAE	0.08
Saudi Arabia	0.05
Kuwait	0.03

Source: Own elaboration based on the data presented in the Summary Table at the end of this chapter

The True Value of Electricity in the UAE

In the Emirates, the price of electricity does not merely reflect the cost of energy but the value of the national energy service: reliable, diversified, clean, and future oriented.

The UAE does not subsidize the present. They invest in the future.

And those who invest with vision — attract the entire world.

Portugal – Environmental Idealism and Nuclear Exclusion

A solar country, but in the shadow of energy pragmatism

Portugal is, in many ways, a unique case in Europe. With enormous solar potential, an extensive Atlantic coast, low population density, and a well-educated population, the country has ideal conditions for a balanced, sovereign, and innovative energy matrix.

However, since the 1970s, Portugal has chosen to exclude nuclear energy from its strategic options, even as many of its neighbors — such as Spain and France — pursued mixed paths combining renewables and nuclear power.

This decision, initially taken out of caution, has evolved into a rigid and ideological political position, remaining unchanged to this day.

Today, Portugal is the only country in Western Europe without any nuclear power plants or an official plan to integrate this technology into its energy mix.

Why did Portugal reject nuclear energy?

The exclusion of nuclear energy in Portugal was based on several factors:

1. Popular fear and political history

The memory of the Chernobyl accident (1986) and later Fukushima (2011) reinforced public fears.

Environmental movements linked to far-left parties, both active and politically influential, cemented an anti-nuclear vision as part of the national ecological identity.

Media narratives influenced by these radical groups prevailed without opportunities for genuine debate.

2. Territorial and demographic dimensions

It has been argued that Portugal is "too small" to accommodate nuclear plants — a questionable argument when compared to examples like Belgium, Slovenia, or Finland.

3. Renewables as a political flag

Since the 2000s, with the liberalization of the electricity sector, Portugal has heavily invested in wind and solar energy, with broad support from the European Union.

The political discourse evolved to claim that "Portugal doesn't need nuclear because it has enough renewables" — a seductive but incomplete view.

Practical consequences of nuclear exclusion

Although the investment in renewables has been successful in several ways, the absence of nuclear power has created significant vulnerabilities:

- High intermittency: Renewable production fluctuates strongly with the weather, requiring imports and natural gas backup.

- External energy dependence: Portugal continues to import electricity from Spain and gas from international markets.

- High consumer prices: Electricity tariffs are among the highest in Europe despite an abundance of sun and wind.

- Persistent residual emissions: Although low, energy sector emissions have not been eliminated due to the continued use of natural gas.

Portugal prides itself on being green — but it depends on what others sell to it.

Chart 62: Evolution of CO_2 Emissions in Portugal

CO₂ Emissions in Portugal (2005-2023)

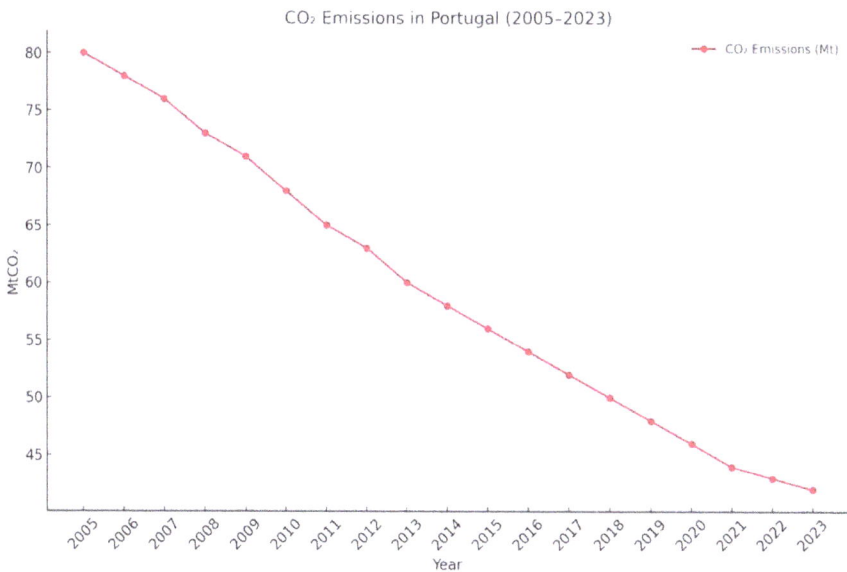

Source: Own elaboration based on the data presented in the Summary Table at the end of this chapter

Graph showing the evolution of CO_2 emissions in Portugal (2005–2023) — illustrating a downward trend, although with residual emissions due to the continued use of natural gas.

Chart 63: Timeline of the Closure of Coal Power Plants

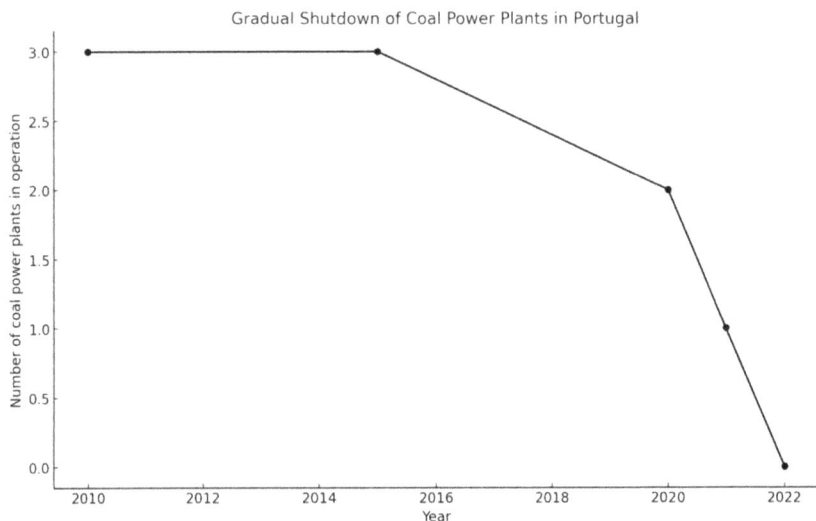

Gradual Shutdown of Coal Power Plants in Portugal

Source: Own elaboration based on the data presented in the Summary Table at the end of this chapter

Graph showing the coal power plant closure plan in Portugal, illustrating the gradual phase-out by 2022.

Key Indicators

Table 70: Key Indicators of Portugal's Energy Policy

Indicator	Current Situation (2023–2024)
CO_2 Emissions	Reduced, but not zero (use of natural gas)
Electricity Mix	60% renewables, 30% natural gas, 10% imports
Nuclear Capacity	None

Energy Imports	High dependence on Spanish and Algerian gas
Electricity Prices	Among the top 5 highest in the EU
Public Acceptance of Nuclear	Low, but growing in technical circles

Source: Own elaboration based on the data presented in the Summary Table at the end of this chapter

The Debate That (Almost) Doesn't Exist

- The Instituto Superior Técnico, the Order of Engineers, and heavy industry entrepreneurs are already advocating for a responsible debate on nuclear energy.

- The 2022 energy crisis, triggered by the war in Ukraine, reignited interest in stable and secure energy sources.

Portugal is a country with the resources, talent, and history to become a balanced energy leader. But the total exclusion of nuclear energy may be limiting that ambition.

Between idealism and pragmatism, Portugal needs a mature debate — not about being nuclear or not, but about being sovereign, sustainable, and fair.

Chart 64: Portugal's Electricity Mix

Portugal Electricity Mix - 2023

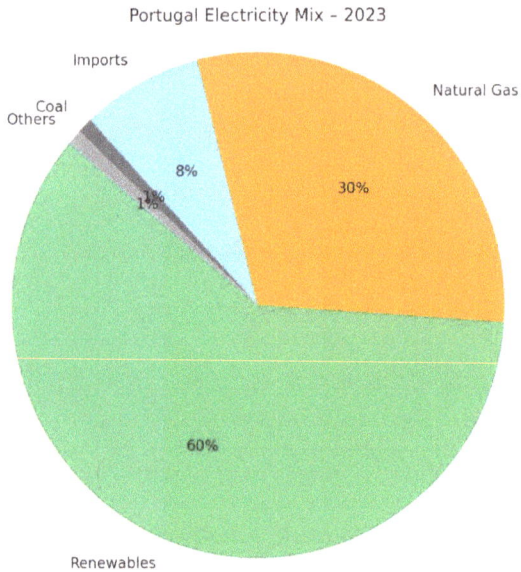

Source: Own elaboration based on the data presented in the Summary Table at the end of this chapter

Graph of Portugal's electricity mix in 2023, showing the predominance of renewable sources but still a strong dependence on natural gas.

Energy Transitions in Portugal

Table 71: Timeline – Milestones of Energy Transitions in Portugal

Year	Energy Milestone
1993	Start of the natural gas network and energy sector liberalization.
1997	Operation of the first international pipeline (Maghreb–Europe Gas Pipeline).
1999	Creation of REN and the start of long-term energy planning.
2003	First large-scale wind farms in operation.
2005	Portugal joins the Kyoto Protocol and boosts investment in renewables.
2007	Major wind expansion with Feed-in Tariffs.
2009	Baixo Sabor Dam was approved under the National Dam Plan.
2010	Start of solar diversification with small-scale PV production.
2012	Full liberalization of the electricity market.
2015	The coal phase-out begins to be scheduled.
2020	Portugal closes almost all coal-fired power plants.

2022	The global energy crisis reignites debate on stable sources and energy security.
2023	The energy mix exceeds 60% of renewables, and there are still no nuclear plans.

Source: Own elaboration based on the data presented in the Summary Table at the end of this chapter

Transition 1: The Arrival of Natural Gas (1990s):

In the late 1980s and early 1990s, Portugal began its first "modern energy transition" with the introduction of natural gas as a large-scale substitute for coal and fuel oil.

Key promoters:

- Luís Mira Amaral (Minister of Industry and Energy under Cavaco Silva).

- Ribeiro da Silva (his energetic Secretary of State), technician and ideologist of energy rationalization.

Objectives:

- Modernize electricity production.

- Reduce emissions of classic pollutants.

- Attract industrial investments with "clean" and predictable energy.

Result:

Portugal became dependent on expensive infrastructure (gas pipelines, LNG terminals, combined cycle plants) and on external suppliers, particularly Algeria and Nigeria.

The grid was modernized, but a new dependence was created — this time fossil and imported.

Chart 65: Evolution of Portugal's Energy Dependence

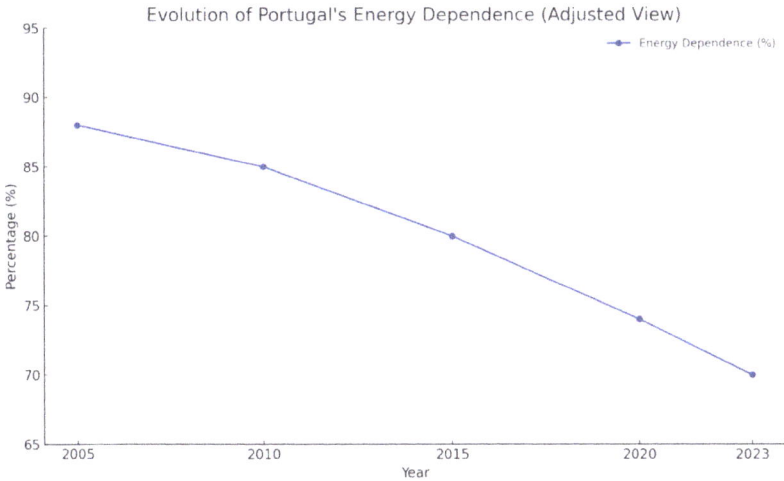

Evolution of Portugal's Energy Dependence (Adjusted View)

Source: Own elaboration based on the data presented in the Summary Table at the end of this chapter

Graph showing the evolution of Portugal's energy dependence, with notable progress over the past two decades — although the country still relies significantly on imports.

Transition 2: The Renewable Boom (2000–2010)

Under the governments of José Sócrates, and later with António Costa as a figure of continuity, Portugal fully embraced the renewable transition — focusing particularly on:

- Large-scale wind power (mainly onshore).

- New-generation hydroelectric dams.

- Solar photovoltaics with incentives and feed-in tariffs.

Political narrative: Portugal would become a world leader in green energy.

Technical reality: Renewable generation did indeed grow significantly — but without solving the problem of intermittency, requiring the continuous maintenance of combined cycle gas plants as a backup.

Chart 66: Evolution of Electricity Prices in Portugal

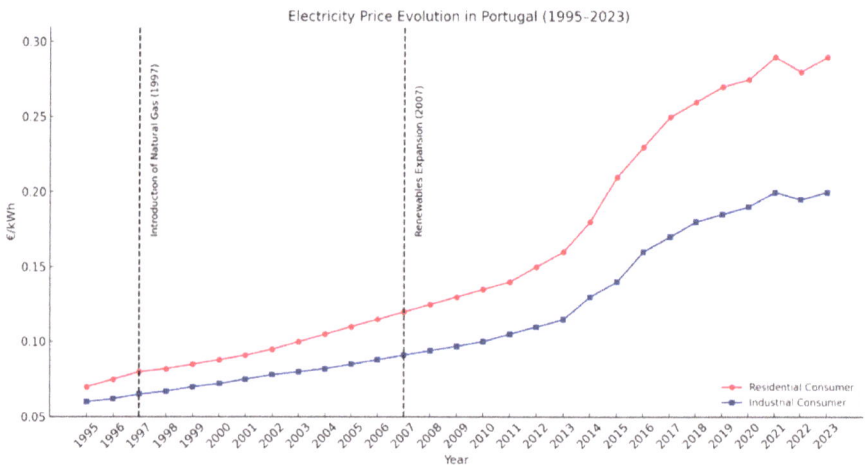

Electricity Price Evolution in Portugal (1995-2023)

Graph showing the evolution of electricity prices in Portugal (1995–2023), highlighting the moments of the two major energy transitions.

Chart 67: Comparative Average Household Electricity Prices between Portugal and Selected EU Partners

Average Household Electricity Prices in the EU (2023)

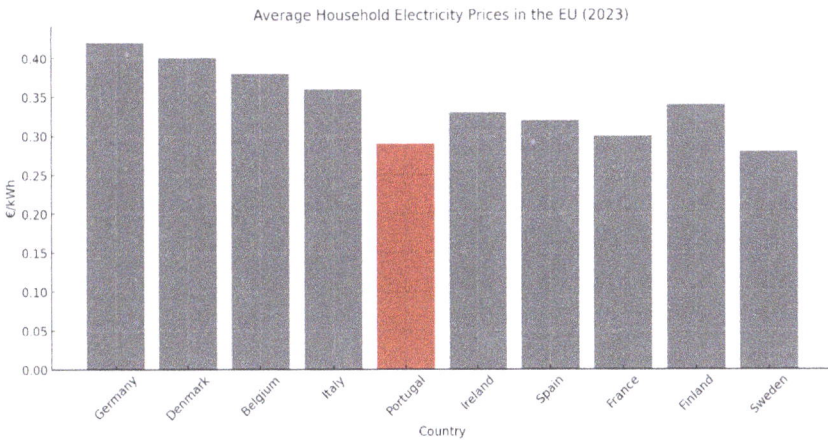

Comparative graph of electricity prices in the European Union (2023), with Portugal highlighted — among the countries with the highest tariffs.

What Was Left Undone?

- Portugal never considered nuclear energy as part of its base solution.

- There is a lack of an energy vision focused on long-term sovereignty and stability.

- The debate became politicized instead of being technical.

Explanation – Feed-in Tariff (FiT) Contracts

Feed-in Tariff (FiT) contracts are energy policy mechanisms used to encourage the production of electricity from renewable sources.

They work as follows: the State or a public operator guarantees the purchase of electricity produced by independent producers (such as wind or solar farms) at a fixed price, generally higher than the market value, for a contractual period (usually between 15 and 25 years.

Main objectives:

- Stimulate private investment in clean and sustainable technologies.

- Reduce market risks for new energy producers.

- Accelerate the adoption of renewable sources before they become competitive on their own.

In Portugal, this model was widely used between 2005 and 2012, primarily to boost wind, hydro, and solar energy.

The contracts were signed with very advantageous tariffs for producers, with the costs indirectly passed on to consumers through electricity bills, grid access tariffs, and general economic interest charges (CIEG).

Chart 68: PPP Comparison of Final Electricity Prices for Consumers in the EU

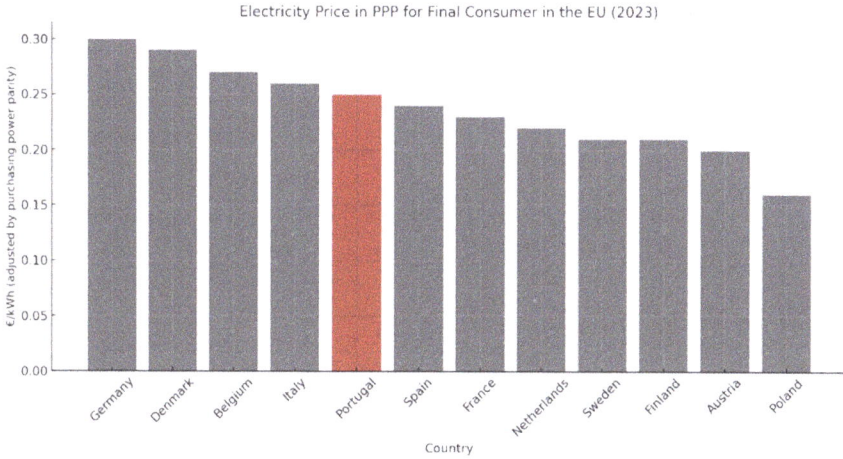

Electricity Price in PPP for Final Consumer in the EU (2023)

Source: Own elaboration based on the data presented in the Summary Table at the end of this chapter

Comparative graph of electricity prices adjusted for Purchasing Power Parity (PPP) in the European Union — with Portugal highlighted.

This graph shows how, even when adjusted for the cost of living, Portugal remains among the countries with the highest prices for final consumers, further reinforcing the critical arguments of our analysis.

What is PPP (Purchasing Power Parity)?

Purchasing Power Parity (PPP) is a method used to adjust the prices of goods and services between different countries, taking into account the local cost of living.

Instead of simply comparing nominal prices (for example, the electricity price in euros per kWh), PPP answers the question:

"What can the average citizen actually buy with their local income?"

Why is PPP important?

Comparing prices without considering purchasing power can be misleading.

The same price can represent a light burden in a wealthy country but a heavy one in a country with lower incomes.

Practical example:

- €0.29/kWh in Portugal and €0.30/kWh in Germany might seem close.

- But the average income in Germany is much higher than in Portugal.

- Result: the real burden of the electricity bill is heavier in Portugal, even if the nominal price is slightly lower.

By comparing electricity prices adjusted for PPP, we can understand the real economic effort that consumers — both domestic and industrial — face in each country.

In summary, PPP allows us to compare countries on "fair terms," revealing where energy is truly more expensive for people's pockets.

What Did the Country Gain from All This?

Positive aspects:

- Significant reduction in CO_2 emissions.

- Internationalization of some companies (e.g., EDP Renováveis).

- Recognition of Portugal's "green brand."

Critical aspects:

- Electricity bills have risen steadily since the 1990s.

- The country did not gain competitiveness — on the contrary, it lost it.

- Grid costs and subsidies were internalized into the final price.

- There was little transparency regarding contracts, guaranteed rents, and overlapping charges (e.g., CIEG, CUST, ERSE, CESE...).

Result?

Portugal became a champion of the transition... and a runner-up in energy prices.

Special Commentary – The Blackout of 28 April 2025: A Harsh Lesson for Portugal

The national blackout of 28 April 2025 brutally exposed the fragility of Portugal's energy system — a direct consequence of decades of irresponsible political decisions.

It is true that the Iberian Peninsula is heavily interconnected internally.

It is also true that the connections between the Iberian Peninsula and the rest of Europe remain scarce and insufficient.

But none of this excuses Portugal's total dependence on Spain's electricity grid and production.

According to reports, the blackout occurred while Portugal was importing 30% of its electricity needs.

And the questions that emerge are devastating:

- Was there no sun in the country that calls itself a "solar powerhouse"?

- Was there no wind in the "European Champion of renewables"?

- Was the drought so severe that hydroelectric dams could not respond?

How is it possible that the nation that boasts about being energy self-sufficient — and even about becoming a net exporter again — collapses at the slightest disturbance in the neighboring country's grid?

The answer is brutal: It was all a lie.

A narrative carefully crafted to favor energy lobbies — and, perhaps, sustained by the entrenched corruption that for decades has marked this sector.

The roots of this fiasco can be clearly traced back to specific political figures:

José Sócrates, who, during his government, launched a reckless and ideologically driven renewable expansion without guaranteeing base load security.

António Costa, who not only perpetuated the same mistakes but also was directly responsible for the premature closure of coal-fired power plants, weakening Portugal's energy resilience — and who now presides over the European Council, a role that should inspire reflection and shame, not promotion.

Meanwhile, who suffers?

The ordinary consumer and the productive sector pay energy prices at the level of rich countries — while Portugal remains among the poorest in the European Union.

The blackout of April 28 is not just an isolated incident.

It is a warning.

A mirror held up to a political class that chose idealistic slogans over serious, sovereign, and sustainable energy policy.

Iberian Peninsula Blackout: A Technical Analysis

Energy Isolation of the Iberian Peninsula

The Iberian Peninsula operates as an 'energy island' within the European electricity system, with limited interconnections to the rest of Europe, especially with France. Currently, the interconnection capacity between Spain and France is only 2%, well below the European target of 15% by 2030.

Probable Causes of the Blackout

The blackout was triggered by a sudden loss of 15,000 megawatts of electricity generation in just five seconds, representing around 60% of Spain's output at that moment. This abrupt drop led to a cascading failure, also affecting Portugal due to the electrical interconnection between the two countries. Although the exact causes are still under investigation, one hypothesis is that an overload of renewable energy production, particularly solar and wind, contributed to the network's instability. The intermittent nature of these energy sources can make it difficult to maintain an immediate balance between generation and demand.

Recovery Capacity: Black Start

Recovery of the power system after a total blackout depends on 'black start' capacity—that is, the ability of certain power plants to restart electricity generation without external grid support. In Portugal, this capacity is limited, with only a few plants, mainly hydroelectric, capable of performing this function. The scarcity of plants with black start capability may have prolonged the time needed to restore electricity supply.

Energy Dependence and Imports

At the time of the blackout, Portugal was importing about 30% of its electricity from Spain, primarily from solar sources. This significant dependency on energy imports, combined with the volatility of renewable sources, may have exacerbated the blackout's effects in Portuguese territory.

Safety Protocols and Interconnections

Electrical interconnections between Portugal and Spain are governed by safety protocols that, in the event of significant voltage or frequency fluctuations, automatically disconnect the systems to prevent major damage. During the blackout, these protection mechanisms were activated, further isolating the two countries' power systems and complicating a quick restoration.

This event highlights the urgent need to strengthen the electrical infrastructure of the Iberian Peninsula, increase interconnection capacity with the rest of Europe, and invest in technologies that ensure greater stability and resilience in the electricity system—especially as renewable sources take on a growing role in the energy mix.

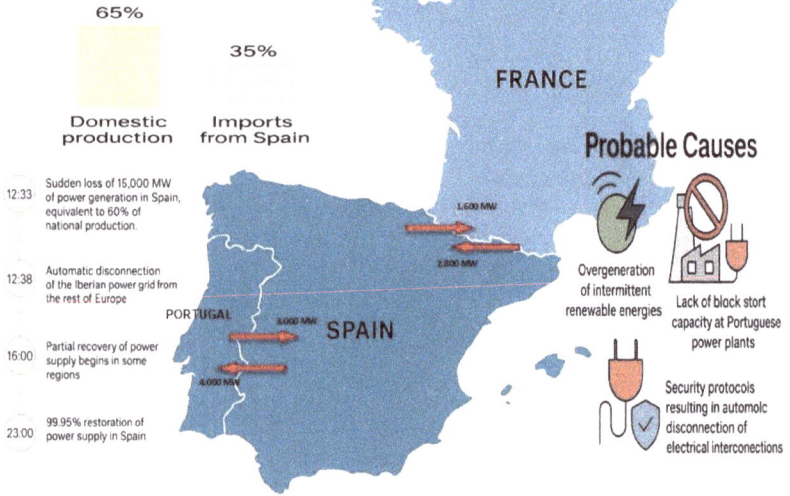

At the time of the blackout Portugal imported more than 30% of its electricity from Spain

65%

35%

Domestic production

Imports from Spain

FRANCE

Probable Causes

12:33 — Sudden loss of 15,000 MW of power generation in Spain, equivalent to 60% of national production.

12:38 — Automatic disconnection of the Iberian power grid from the rest of Europe

16:00 — Partial recovery of power supply begins in some regions

23:00 — 99.95% restoration of power supply in Spain

1.600 MW

2.800 MW

PORTUGAL   3.000 MW   SPAIN

4.000 MW

Overgeneration of intermittent renewable energies

Lack of block stort capacity at Portuguese power plants

Security protocols resulting in automoic disconnection of electrical interconections

Comparative Analysis – Four Strategies, Four Outcomes

Comparative Table – Key Indicators

Table 72: Comparative Table – Energy Strategies and Outcomes

Indicator	Germany	Finland	UAE	Portugal
Nuclear in the energy mix	0% (shut down)	30%	25%	0%
CO_2 emissions (per capita)	High	Moderate	Moderate-low	Moderate

416

Electricity price (PPP)	Very high	High	Moderate	High
External dependence	High	Reduced	Low	High
Grid stability	Good	Excellent	Excellent	Good (intermittent)
Public acceptance of nuclear	Low	High	High	Low
Long-term strategic vision	Inconsistent	Clear and steady	Ambitious and coherent	Idealistic, no technical base

Source: Own elaboration based on the data presented in the Summary Table at the end of this chapter

Note: Approximate data based on official sources (Eurostat, WNA, IEA, REN, ERSE, FANR).

Strengths and Weaknesses of Each Model

Germany

- Strength: Capacity for mobilization and investment in renewables.

- Weakness: Abandonment of nuclear energy without ensuring energy stability — resulting in dependence on coal, gas, and high prices.

Finland

- Strength: Balance between nuclear and renewables, energy sovereignty, and investment in new technologies.

- Weakness: High cost of living and geographic isolation, which limit interconnections.

United Arab Emirates

- Strength: Long-term planning, exemplary execution, energy independence, and regional leadership.

- Weakness: Higher tariffs in the Gulf context (but sustainable) and the future challenge of industrial diversification.

Portugal

- Strength: High share of renewables and strong environmental performance.

- Weakness: High electricity prices, external dependence, and lack of a stable base for production.

Chart 69: Comparison Between the Various Energy Models

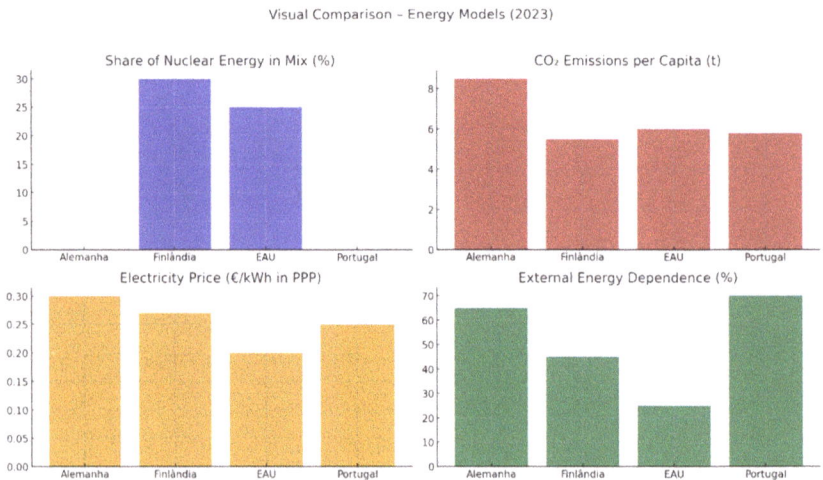

Visual Comparison – Energy Models (2023)

Source: Own elaboration based on the data presented in the Summary Table at the end of this chapter

Comparative visual infographic with the four energy models (Germany, Finland, UAE, and Portugal), representing:

• Share of nuclear energy in the mix,

• CO_2 emissions per capita,

• Electricity price adjusted by PPP,

• Level of external energy dependence.

Chart 70: Energy Effort: Price vs. Per Capita Income

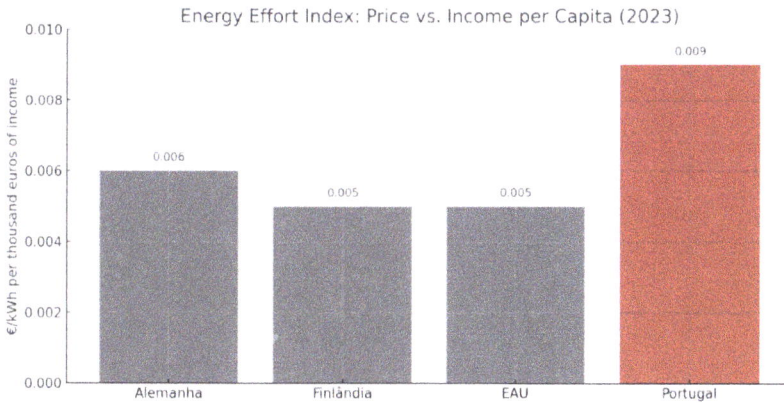

Energy Effort Index: Price vs. Income per Capita (2023)

Source: Own elaboration based on the data presented in the Summary Table at the end of this chapter

Graph of the Energy Effort Index, relating the adjusted electricity price (PPP) to the per capita income of each country.

This graph shows how much 1 kWh weighs on citizens' wallets, in proportion to their income — and reveals that, proportionally, Portugal is the country that "suffers" the most from electricity costs among the four compared.

Chart 71: GDP Growth vs. Electricity Price

GDP Growth vs. Electricity Price (last 10 years)

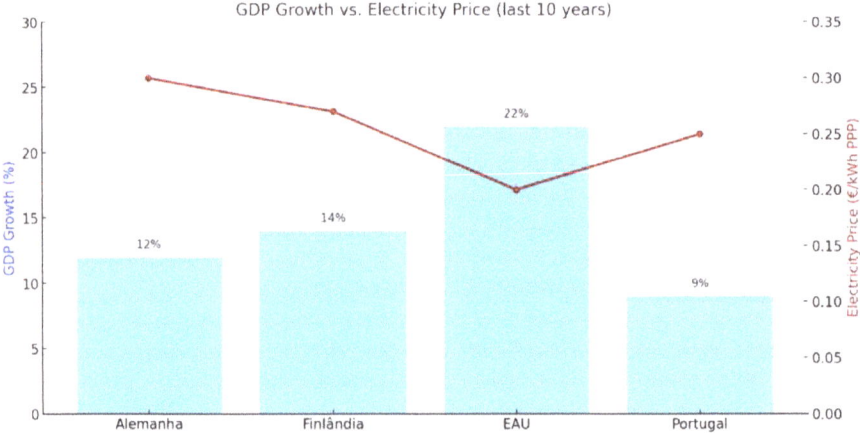

Source: Own elaboration based on the data presented in the Summary Table at the end of this chapter

Comparative graph of GDP growth over the past 10 years vs. electricity price (adjusted by PPP) for the four countries analyzed.

This graph clearly shows that countries with a more strategic energy vision (such as the UAE) managed to achieve greater growth while maintaining lower energy prices, whereas others — like Portugal and Germany — faced modest growth coupled with high tariffs.

And yes, with due caution, it is possible to establish a plausible and well-founded relationship between lower economic growth and high electricity prices, especially in countries with high energy dependence and reduced industrial maneuverability.

Expensive Energy, Limited Growth?

The comparison between electricity prices and economic growth over the past 10 years suggests a worrying pattern:

Countries with high electricity prices — adjusted by purchasing power parity — recorded more modest economic growth during the analyzed period.

This is particularly evident in:

- Portugal, with a cumulative growth of only 9% and one of the highest relative energy costs in the EU.

- Germany, which faced timid growth (12%) despite being one of the most robust economies — was burdened by inflated electricity bills caused by the nuclear phase-out and gas dependence.

In contrast:

- The United Arab Emirates, with abundant, diversified, and well-managed energy, grew 22% over the same period.

- Finland, with a balanced model between renewables and nuclear, also maintained growth above the Eurozone average.

Of course, the economy depends on multiple factors — such as demographics, fiscal policy, innovation, and exports — but energy is a transversal and decisive factor, especially:

- For heavy and manufacturing industries,

- For national production costs,

- For the country's attractiveness to foreign investment,

- And for households' disposable income.

When energy is expensive, everything else becomes more costly — and growth becomes more difficult.

Thus, although we cannot affirm an absolute direct causality, the data strongly supports a correlation between affordable energy and sustained economic growth.

Denying this relationship would be ignoring the physical foundations of the economy.

The Dimension of Time: Short-Term Thinking or Long-Term Vision?

A fundamental lesson from this comparison is that energy decisions must be evaluated over the long term.

The most "modern" or "popular" solution is not always the most effective one.

Countries that invested in continuity, stability, and technological diversification (like Finland and the UAE) are now reaping more robust results.

Energy transitions based on ideology — rather than science and engineering — tend to be costly, both economically and geopolitically.

Conclusion of the Present Chapter – Decisions That Cost, Strategies That Pay Off

In this chapter, we explored four profoundly different energy trajectories:

- **Germany**, which abandoned nuclear power for political reasons and reverted to coal and gas.

- **Finland**, which combined science, strategy, and technology to ensure sovereignty and stability.

- The **United Arab Emirates**, which invested with vision and pragmatism, even without a nuclear tradition.

- **Portugal**, which bet everything on renewables without a solid technical base — and without a Plan B.

The comparison of their results reveals something obvious yet often forgotten:

Energy is not an ideology — it is infrastructure, competitiveness, and sovereignty.

When energy decisions are based on:

- Emotions,

- Media pressures,

- Electoral calculations,

- Or convictions not supported by data,

... the results inevitably impact household budgets, corporate costs, and the limited growth of nations.

The Price of Well-Intentioned but Poorly Planned Actions

Portugal and Germany are examples of energy idealism without a solid strategy.

Electricity prices have become among the highest in Europe, and external dependence remains.

Green energy paradoxically became a factor of inequality rather than a driver of accessible progress.

The Strength of Informed Vision

Finland and the UAE, on the other hand, demonstrate that:

- A small country or one with a harsh climate can also lead — if it studies and plans carefully.

- Civil nuclear power, when well-managed, can coexist with renewables and make them more viable.

- Technical, rather than ideological, decisions create more sustainable — and fairer — societies.

Cheap and clean energy is possible — but it demands political courage, technical intelligence, and humility before the facts.

To the Reader

The message we leave is simple:

- Studying matters.

- Comparing matters.

- Questioning also matters.

Those who aspire to a better, fairer, and more sustainable country...

... should begin by asking:

"*What are we basing our energy decisions on?*"

Table 73: Sources Consulted in Chapter 8

Source	Description	Usage in Chapter
Eurostat	European Union statistical agency	CO_2 emissions data, electricity prices, and energy dependence
IEA – International Energy Agency	Global energy agency	National energy profiles, matrix data, projections
World Nuclear Association (WNA)	Global Nuclear Energy Association	Nuclear capacity and project data for the analyzed countries
REN – Redes Energéticas Nacionais (Portugal)	Portugal's grid operator	Electricity matrix and Portuguese energy transition data
ERSE – Energy Services Regulatory Authority (Portugal)	Portuguese energy and gas regulator	Electricity tariffs and price evolution in Portugal
Destatis (Germany)	German National Statistics Office	Macroeconomic and energy data for Germany
BMWK (Germany)	Ministry of Economy and Climate Protection	German energy policy, nuclear phase-out, and renewables
FANR (UAE)	Federal Authority for Nuclear Regulation (UAE)	Barakah project and nuclear licensing data

UAE Energy Strategy 2050	UAE National Energy Strategy document	Energy planning and nuclear/renewable targets
Statistics Finland	Finnish National Statistics Office	Energy matrix, economic growth, and emissions data
VTT Technical Research Centre of Finland	Finnish Technological Research Center	Studies on nuclear, renewables, and energy security
DGEG – Portugal	Portuguese energy authority	Historical data and energy sector evolution
IAEA	International Atomic Energy Agency	Global nuclear capacity and regulatory data
EPOV	European Energy Poverty Observatory	Impact indicators of electricity prices on domestic consumption
World Bank	International financial institution	GDP and economic growth data
OECD	Organization for Economic Cooperation and Development	Income, electricity prices, and development data

Preparation for the Next Chapter – Energy Transition and the Role of Rare Earths

The world is undergoing a profound and inevitable transformation: the shift from a fossil fuel-based energy model to a cleaner, more resilient, and sustainable system.

This process, known as the energy transition, is not merely a technological change — it is a civilizational restructuring involving economics, politics, geopolitics, science, security, and ecology.

In this journey, nuclear energy reemerges as a strategic pillar.

With low carbon emissions and high energy density, nuclear power has the potential to provide a stable large-scale electricity supply, complementing intermittent sources like solar and wind.

However, to achieve a truly effective energy transition, another often overlooked factor must be brought to the forefront: **critical minerals and rare earth elements**.

In the next chapter, we will explore:

- What the energy transition really is and why it is so urgent.
- The crucial role of nuclear energy in this transformation.
- The strategic dependence on critical minerals and rare earths for the production of clean technologies.

- The geopolitical, environmental, and social challenges related to the extraction, processing, and supply of these resources.

- And finally, the interconnection between energy sovereignty, climate security, and control of global supply chains.

The energy transition is not just a race for clean technology.

It is a new gold rush — but this time, with isotopes, rare earths, and megawatts at the heart of the contest.

Chapter 9 –
Energy Transition and the Role of Rare Earths

This chapter will connect the need to replace fossil fuels with the crucial role of critical minerals and rare earth elements in modern energy technology.

What is Energy Transition?

The energy transition is a structural process of change in how humanity produces, distributes, and consumes energy. It involves a progressive shift from a fossil fuel-based model – such as coal, oil, and natural gas – to a more sustainable, diversified, and low-carbon energy system, incorporating renewable sources, electrification, and energy efficiency.

Why is the Energy Transition Necessary?

The need for an effective energy transition arises from three major global challenges:

- Climate Crisis – The burning of fossil fuels is the main driver of increased greenhouse gas (GHG) concentrations in the atmosphere, causing global warming and extreme weather events.

- Energy Security – Dependence on fossil fuels makes countries vulnerable to geopolitical shocks and energy price fluctuations.

- Resource Sustainability – Coal, oil, and gas are finite resources, and their extraction has severe environmental impacts.

The Main Pillars of Energy Transition

The energy transition is not a single or linear process but involves various simultaneous strategies:

1. Decarbonization – Reducing CO_2 emissions by replacing fossil sources with clean energy.

2. Electrification – Expanding electricity use in sectors traditionally reliant on fossil fuels, such as transport and heating.

3. Energy Efficiency – Improving the performance of energy systems to reduce waste and optimize demand.

4. Diversification of the Energy Matrix – Integrating diverse sources to reduce dependence on a single type of energy.

5. Storage and Smart Grids – Developing advanced batteries and energy management systems to address the intermittency of renewables.

Different Approaches to Energy Transition

Not all countries follow the same path in the energy transition. Approaches vary according to available natural resources, energy policies, and socioeconomic challenges:

- **European Union** – Adopted aggressive decarbonization and renewable expansion targets, aiming for carbon neutrality by 2050.

- **United States** – A mix of renewable sources and investment in new nuclear technologies.

- **China** – Leads in renewables but continues heavy use of coal.

- **Middle Eastern Countries** – Despite reliance on oil, they are investing in solar energy and green hydrogen.

Energy Transition: Gradual or Disruptive?

The transition can occur in two main ways:

• Gradual – Incremental and progressive changes that ensure energy stability but may be slower.

• Disruptive – Abrupt transformations driven by crises or disruptive technological advances.

Both approaches present challenges. Gradual change may be too slow to mitigate climate impacts, while disruptions can cause economic and social instability.

Energy transition is one of the greatest challenges of the 21st century. It requires massive investments, technological innovation, and global cooperation. No single energy source alone will solve the sustainability equation – a balanced mix of renewables, nuclear, and new storage technologies will be essential.

As we move forward in this chapter, we will explore how nuclear energy, and critical minerals play a vital role in this transformation.

Nuclear Energy and Energy Transition

Nuclear energy has been one of the most debated topics within the energy transition. While some countries have reduced their nuclear share, others continue to expand it as a reliable, low-carbon alternative to ensure energy security and reduce greenhouse gas emissions.

Expansion of Nuclear Energy in Energy Transition

Despite challenges such as high upfront costs and concerns about nuclear waste, many nations are investing in the nuclear sector as a solution to decarbonize their economies. In recent years, countries like France, China, Russia, and India have significantly expanded their nuclear capacity, while the United States and the United Kingdom are working to revitalize their nuclear infrastructure through new technologies.

Strategic Advantages of Nuclear Energy

- High Energy Density – A small amount of nuclear fuel generates large amounts of electricity.

- Continuous Operation – Unlike intermittent sources such as solar and wind, nuclear provides constant electricity.

- Smaller Land Footprint – Nuclear reactors require less space than solar or wind farms to produce the same amount of energy.

- Geopolitical Resilience – Reduces reliance on imported fossil fuels, enhancing energy security.

New Technologies and the Future of the Nuclear Sector

Technological advances are shaping the future of nuclear energy. Small Modular Reactors (SMRs) promise greater safety, flexibility, and lower costs, while advanced research on thorium reactors and nuclear fusion aims to deliver more sustainable and safer alternatives.

The Urgency for Exponential Growth in Nuclear Energy

Data from the past 20 years show that renewable sources such as solar and wind have grown exponentially, while nuclear energy has grown much more modestly. This creates a critical challenge: although renewables are essential for the energy transition, their intermittency and low energy density make an energy matrix based solely on these sources insufficient.

Nuclear energy, on the other hand, offers stable electricity generation, low carbon emissions, and high energy density, making it an essential component for meeting decarbonization goals. Without accelerated expansion of the nuclear sector, global climate objectives are at serious risk, as relying exclusively on renewables would mean dealing with issues related to energy storage, infrastructure, and demand response capacity.

Countries that have made the most progress in the energy transition and reduced their CO_2 emissions—such as France and Sweden—have done so thanks to the strong presence of nuclear energy in their electricity mix. The stagnation of the nuclear sector in various regions over the past decades has hindered decarbonization efforts. Unless nuclear growth keeps pace with that of renewables, achieving carbon neutrality by 2050 will become an unattainable goal.

Therefore, to ensure a secure, sustainable, and carbon-free energy future, it is essential that nuclear energy grows at a rate comparable to that observed for solar and wind in recent decades. Emerging technologies such as SMRs (Small Modular Reactors), advanced thorium reactors, and nuclear fusion can play a key role in this process.

The energy transition cannot rely solely on intermittent solutions. Complementarity between renewables and nuclear is the only viable strategy to guarantee reliable, clean, and affordable electricity for all.

Chart 72: Global Energy Mix – 2023

Global Energy Mix - 2023

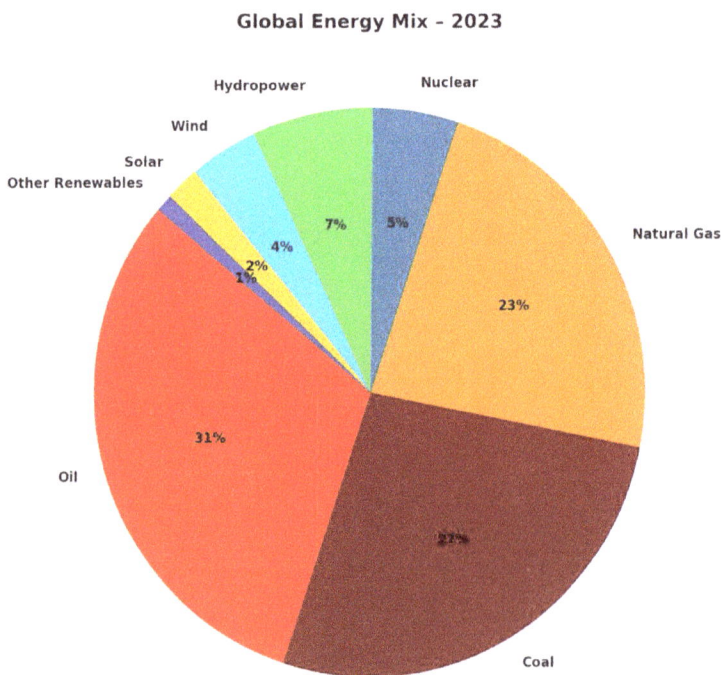

Source: Own elaboration based on the data presented in the Summary Table at the end of this chapter

Chart 73: Growth of Low-Carbon Energy Sources (2003–2023)

Growth of Low-Carbon Energy Sources (2003-2023)

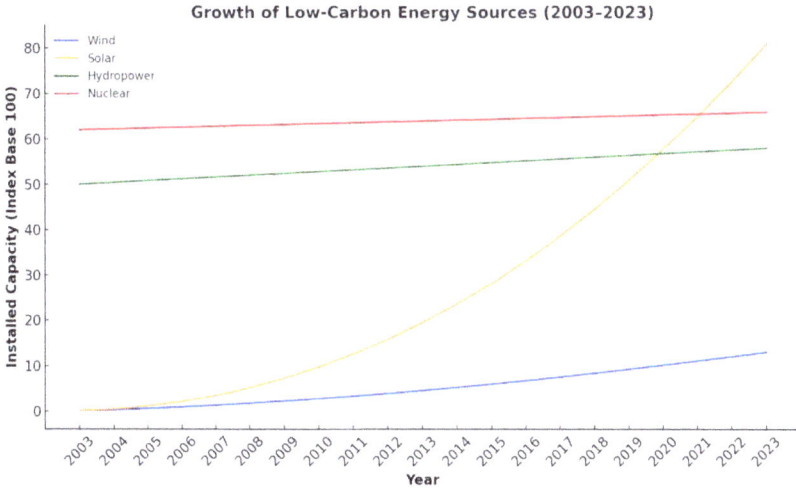

Source: Own elaboration based on the data presented in the Summary Table at the end of this chapter

The data from the chart projecting global economic growth and the corresponding energy needs make it clear that the only way to meet the rising energy demand without compromising carbon neutrality goals is through a massive expansion of nuclear energy.

Chart 74: Projection of Economic Growth, Energy Demand, and Carbon Neutrality (2020–2050)

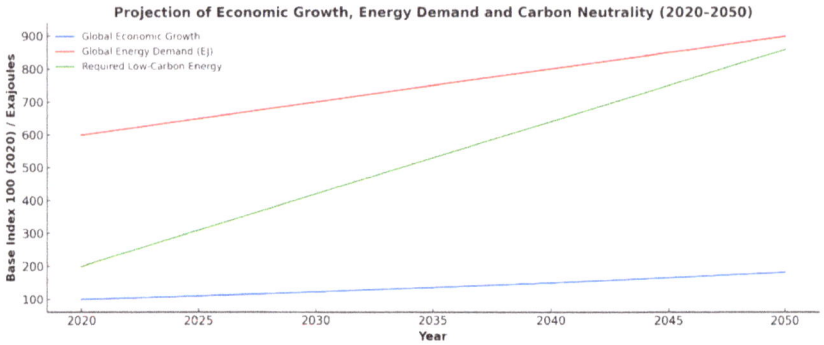

Projection of Economic Growth, Energy Demand and Carbon Neutrality (2020-2050)

Source: Own elaboration based on the data presented in the Summary Table at the end of this chapter

The exponential growth of renewable sources such as solar and wind is crucial, but their intermittency and reliance on weather conditions limit their ability to provide electricity reliably and continuously. To ensure stability in the global power system and enable sustainable economic development, nuclear energy must keep pace with—or even exceed—the growth rate of renewables.

If nuclear capacity continues to grow slowly, the world will be forced to rely on natural gas and other fossil sources to bridge the gap between supply and demand, which would severely undermine decarbonization efforts. The experience of countries like France and Sweden, which drastically reduced their emissions through intensive use of nuclear energy, proves that this is the most effective path toward a clean and resilient energy system.

The accelerated expansion of nuclear energy is the only viable alternative to ensure a sustainable energy future capable of meeting global demand without compromising climate goals.

Nuclear energy can play a crucial role in the energy transition, especially as a reliable complement to renewables. Despite the challenges, technological advancements indicate that the sector can grow safely and sustainably in the coming years.

The Need for Critical Minerals and Rare Earths

Critical minerals and rare earth elements are essential to the energy transition. These materials are fundamental to produce solar panels, wind turbines, lithium-ion batteries, advanced nuclear reactors, and various other energy technologies. Without these raw materials, the advancement of clean energy would be severely compromised.

Geopolitics of Critical Minerals

Critical minerals—including rare earths, lithium, cobalt, nickel, and graphite—have become vital for the energy transition and strategic industries such as transportation electrification, energy storage, and semiconductor manufacturing. However, their production and refining are highly concentrated in a small number of countries.

- **China**: Holds about 60–70% of global rare earth extraction and over 85% of refining capacity. It also leads in graphite refining and dominates the lithium-ion battery supply chain.

- **Democratic Republic of Congo (DRC):** Produces about 70% of the world's cobalt, a mineral essential for advanced batteries.

- **Indonesia and the Philippines**: Major producers of nickel, key for manufacturing high-performance batteries.

- **Australia and Chile**: Together, they control the majority of lithium production, essential for electric vehicle batteries.

- **Russia**: A key supplier of nickel and palladium, both essential for various industrial and technological applications.

This concentration of production and refining in just a few countries creates geopolitical risks and strategic vulnerabilities for the West.

An infographic showing the main rare earth deposits around the world.

Table 74: Leading Countries with Significant Rare Earth Deposit

Country	Estimated Reserves (million tons)	Key Notes
China	44	World's largest producer and refiner; dominates the value chain.
Vietnam	22	Vast reserves and increasing production in recent years.
Brazil	21	High potential deposits in Amazonas state and Minas Gerais.
Russia	19	Significant reserves; mining affected by sanctions.
India	6.9	Coastal reserves; production is still limited.
Australia	4.2	Strong exporter; Mount Weld mine is a highlight.
United States	2.3	Mountain Pass mine (California); recent focus on industrial revival.
Canada	2.0 (estimated)	Untapped potential; projects under environmental review.

Western Dependence on China and Risks to the Supply Chain

China's dominance in the extraction and refining of critical minerals is not just an economic advantage but also a powerful geopolitical tool. The West is heavily dependent on Chinese supplies for industries such as:

- Electric vehicles (EVs)

- Wind turbines

- Solar panels

- Semiconductors and advanced electronic equipment

- Defense and aerospace technology

The risks of this dependence include:

Price and access manipulation: China has already restricted exports of strategic elements like gallium and germanium to pressure its geopolitical rivals.

Trade blockades and sanctions: In a scenario of global conflict or trade tensions, Beijing could use control over these minerals as diplomatic leverage.

Lack of alternative refining capacity: Even if the West extracts these minerals in allied countries, the lack of local refining infrastructure remains a critical bottleneck.

The United States, the European Union, and other nations are increasingly concerned about these risks and are seeking ways to reduce this vulnerability.

Chart 75: Leading Rare Earth Producers (2023)

Top Rare Earth Producers (2023)

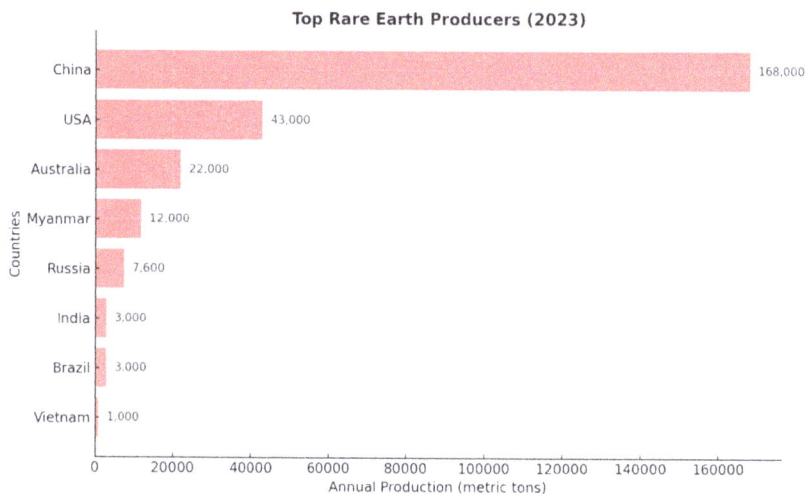

Source: Own elaboration based on the data presented in the Summary Table at the end of this chapter

Table 75: Major Rare Earth Producers (2023)

Country	Global Share (%)
China	70%
United States	14%
Australia	6%
Myanmar	5%
Others	5%

Source: Own elaboration based on the data presented in the Summary Table at the end of this chapter

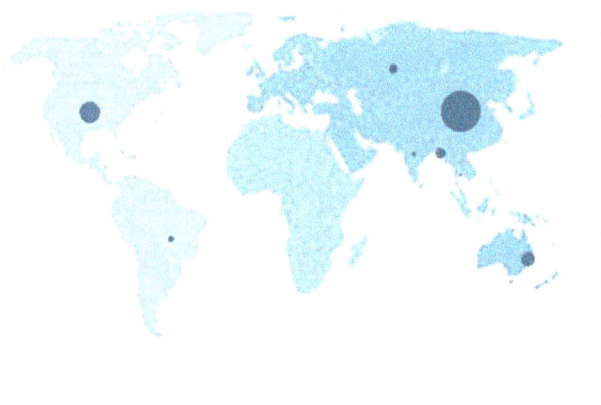

Map showing the main countries with rare earth production capacity. China overwhelmingly dominates the sector, with about 85% of global capacity, followed by Malaysia, the USA, Australia, and a small refining hub in Estonia.

Source: Own elaboration based on the data presented in the Summary Table at the end of this chapter

Strategies to Reduce This Dependence

Given this strategic vulnerability, various initiatives are being developed to diversify the supply chain and reduce dependence on China. The main strategies include:

1. **Responsible Mining and Expansion of Production in Allied Countries**

   - Encouraging exploration and extraction of rare earths in countries such as Australia, Canada, Brazil, and the USA.

   - Development of balanced environmental regulations that allow sustainable extraction without major ecological impacts.

2. **Investment in Mineral Refining and Processing Outside China**

   - Construction of refining plants in Europe and North America to decrease Chinese dependency.

   - Strategic partnerships with countries like Japan and South Korea, which already have expertise in refining certain minerals.

3. **Recycling of Critical Minerals**

   - Programs to recover rare earth elements and valuable metals from used batteries, discarded electronic equipment and decommissioned wind turbines.

   - Technological advances to make recycling more efficient and cost-effective.

4. New Sources of Extraction

- Deep-sea mining, exploring polymetallic nodules on the ocean floor, although with significant environmental challenges.

- Extraction of minerals from alternative sources, such as old mining tailings and new geological formations.

5. Geopolitical Agreements and Alliances

- Creation of alternative supply blocks, such as the Mineral Security Partnership (MSP) led by the USA.

- Bilateral agreements between the EU and producing countries to ensure a stable supply of strategic materials.

Table 76: Strategies to Reduce Dependence on China

Strategy	Example	Objective
Supplier diversification	Australia, Brazil, Canada	Reduce geographic concentration
Material recycling	EU, Japan	Circular economy and less extraction
Investment in substitutes	MIT, European startups	Non-critical alternative materials

Source: Own elaboration based on the data presented in the Summary Table at the end of this chapter

What Are Rare Earths and Why Are They Essential?

Rare Earths are a group of 17 chemical elements in the Periodic Table, including lanthanum, neodymium, terbium, and dysprosium, among others. Despite the name, these elements

444

are not exactly 'rare' in the Earth's crust but are dispersed in low concentrations, making their extraction and separation difficult and costly.

Why Are They Important?

Rare earths are essential to modern technology due to their unique magnetic, optical, and electrical properties. They play an irreplaceable role in several strategic industries, including:

1. Energy Transition

- Neodymium magnets are used in wind turbines and electric vehicle (EV) motors.

- Dysprosium improves these magnets' resistance to high temperatures.

2. Electronics and Semiconductors

- Smartphones, laptops, and television screens use rare earth phosphors to display vivid colors.

- Chips and electronic components rely on these materials for conductivity and miniaturization.

3. Defense and Military Technology

- Advanced sensors, radar systems, and missile guidance systems use rare earths.

- The F-35 and other modern military aircraft depend on materials based on these elements.

4. Medicine and Healthcare

- Gadolinium is used in magnetic resonance imaging (MRI) to provide clearer images.

The Chinese Monopoly and the Global Race

China controls about 85% of the world's rare earth refining and uses this dominance as a geopolitical tool. The West and other powers are rushing to develop alternatives, including new mines, recycling, and strategic agreements with resource-rich countries like Australia, Brazil, and Canada.

Rare earths are not just common materials—they are the building blocks of modern technology. Without them, the energy transition, the digital era, and the advancement of the defense industry would be impossible. The growing competition for their control may redefine the global balance of power in the coming decades.

Table 77: Global Rare Earth Refining Capacity

Country	Refining Capacity (%)
China	85%
Malaysia	6%
USA	5%
Australia	2%
Estonia	2%

Source: Own elaboration based on the data presented in the Summary Table at the end of this chapter

Geological Resources in Europe

Europe does not have significant rare earth production, but that does not mean these minerals are absent from the continent. The issue is a combination of geological, political, environmental, and strategic factors. Let's break it down:

1. Geological Resources in Europe

Europe has rare earth deposits, but they are smaller and less accessible compared to other regions. Some examples include:

- **Norway and Sweden:** Have rare earth deposits, particularly in northern Scandinavia.

- **Greenland (Denmark):** Rich reserves of rare earths have been discovered, but extraction faces environmental and political barriers.

- **France and Spain**: Small deposits have been identified but are not commercially explored.

The reality is that, although these resources exist, the richest and most economically viable deposits are found in China, Australia, and the Americas.

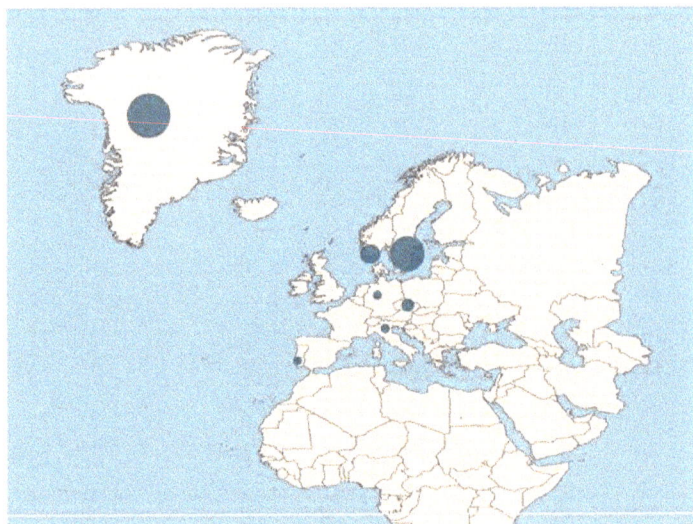

Map highlighting the main rare earth deposits in Europe.

Note: In Italy and Portugal, only traces have been detected so far, with no estimated volumes.

Table 78: Main Rare Earth Deposits in Europe

Location	Country	Estimated Volume (Mt REO)	Current Status
Kvanefjeld (Ilimaussaq)	Greenland	~1.5	One of the largest in the world,

			politically sensitive.
Kiruna (Per Geijer)	Sweden	>1.0	Discovered in 2023, it is considered the largest in the EU.
Norra Kärr	Sweden	~0.6	High potential; facing environmental opposition.
Tanbreez	Greenland	>1.0	Private project, rich in heavy rare earths.
Fen Complex	Norway	~0.5	Under development: carbonatite rocks.
Tisová	Czechia	~0.2	Under assessment, ore associated with by-products.
Krásno and Mariánské Lázně	Czechia	—	Preliminary exploration.
Storkwitz	Germany	~0.03	Known deposit; environmentally sensitive.
Piedmont	Italy	—	Traces of reserves in alpine regions.
Serra de Monchique (traces)	Portugal	—	Sporadic occurrences; no active exploration.

2. Political and Environmental Reasons for the Lack of Production

The main reason why Europe does not produce rare earths is not a lack of minerals but rather strict environmental policies and high extraction and refining costs.

Restrictive Environmental Legislation

- The process of mining and refining rare earths generates highly toxic and radioactive waste.

- Europe enforces stringent environmental regulations that make extraction expensive and bureaucratic.

- Pressure from environmental groups hinders the opening of new mines.

Strategic Dependence on China and Other Countries

- For decades, Europe opted to import rare earths from China, as it was cheaper and less environmentally problematic.

- However, due to rising geopolitical tensions and growing tech competition, the EU now sees this dependence as a strategic risk.

High Extraction Costs in Europe

- In addition to regulations, the costs of extracting and refining in Europe are higher than in China or Australia.

- The investment required to make production economically viable has not paid off until recently.

3. The New European Race for Critical Minerals

With increasing rivalry between the West and China, the European Union is shifting course and investing in rare earth mining. Some initiatives include:

Norwegian and Swedish Projects for Rare Earth Mining

- Sweden announced in 2023 the discovery of the largest rare earth deposit in Europe.

- Norway is developing technologies for sustainable mining.

EU Strategic Partnerships with Mining Countries

- The EU is pursuing agreements with Canada, Australia, and Brazil to secure a stable supply without depending on China.

Investment in Critical Mineral Recycling

- Europe is investing in the recovery of rare earths from discarded batteries and electronic equipment as an alternative to traditional mining.

Europe Once Again Falling Behind in the Technology Race?

The energy transition and digital revolution have triggered a new global competition for so-called critical minerals, among which rare earths are the most strategic. Despite having significant deposits, Europe fell behind in mining and refining these materials essential to EVs, wind turbines, semiconductors, and defense technologies.

The European Reality: Resources Exist, but Strategy and Political Will Are Lacking

Contrary to the narrative that Europe lacks rare earths, important deposits exist in Sweden, Norway, Greenland, France, and Spain. The issue lies not in geology but in environmental barriers, bureaucracy, and excessive dependence on China.

- China Dominates Global Refining

    - Even if Europe extracted its own rare earths, it lacks refining capacity. Currently, 85% of global processing is done in China.

- Delay in Strategic Decisions

    - While the US, Australia, and Canada are rapidly expanding their supply chains, the EU remains slow and divided on how to proceed.

- Environmental Challenges and Bureaucracy

    - The extraction and refining process generates toxic waste, and EU environmental laws make new projects difficult.

    - Environmentalist groups block initiatives, even when more sustainable solutions are proposed.

Sweden and Norway: Europe's New Hope?

In 2023, Sweden announced the largest rare earth discovery in Europe near Kiruna. The country now aims to fast-track exploration, but experts warn that actual production may still be 10 to 15 years away. Norway is planning land and deep-sea mining but faces environmental resistance.

Meanwhile, Europe remains highly dependent on China, leaving its industry vulnerable to trade restrictions and geopolitical instability. Without a rapid shift in strategy, the continent risks falling behind once more in the race for new technologies and the energy transition.

Will Europe Produce Rare Earths in the Future?

Yes, Europe is reassessing its position and seeking ways to initiate local production, but it will still take years to become competitive. The environmental issue remains a major challenge, and the European strategy is more focused on recycling and trade agreements than on large-scale mining.

If Europe wants to secure its technological and energy independence, it will need to accelerate its efforts to develop its own sources of rare earths. Otherwise, it will remain vulnerable to market manipulation and Chinese restrictions.

Controversy has arisen in Europe regarding President Trump's intention to acquire Greenland, which remains under Danish administration. It now seems clear that the US's intention was to protect and exploit the island's potentially vast rare earth and critical mineral reserves.

The argument of protecting US national security may imply that Europeans will do nothing with the resources found there—or worse, they could fall under Chinese or Russian influence.

The lack of action—or even inaction—by Europeans and the EU on these strategic matters is prompting the US to move decisively to protect interests that ultimately also affect Europe.

The EU, still clinging to its 'political correctness' mindset and unwilling to embrace disruption or assertive power, continues to be conditioned by outdated environmentalist narratives that solve nothing. It risks being surpassed by nations with clear visions and concrete strategies.

We may very well see Greenland become the 51st state of the United States—perhaps to the delight of its 70,000 inhabitants.

Table 79: Everyday Devices that Use Rare Earths and Critical Metals

Equipment	Rare Earths Used	Critical Metals Used
Smartphones	Neodymium, Europium, Terbium, Yttrium	Lithium, Nickel, Cobalt
Electric Vehicles	Neodymium, Dysprosium, Lanthanum	Lithium, Nickel, Cobalt, Graphite
Wind Turbines	Neodymium, Dysprosium, Praseodymium	Nickel, Cobalt
Computers and Laptops	Yttrium, Europium, Terbium	Lithium, Copper
TVs and Monitors	Yttrium, Europium, Terbium	Copper, Aluminum

Headphones and Speakers	Neodymium, Dysprosium	Copper, Aluminum

Source: Own elaboration based on the data presented in the Summary Table at the end of this chapter

These examples show that almost everything we use in our daily lives contains some amount of rare earth elements, making them indispensable to modern technology.

The geopolitics of critical minerals has become a key factor in the struggle for technological and energy dominance in the 21st century. The energy transition and the digital revolution increasingly rely on resilient and diversified supply chains. If the West does not reduce its dependence on China, it could face significant strategic vulnerabilities in the coming decades.

Environmental and Social Impacts of Mineral Extraction

The extraction of rare earth elements and critical metals is essential for the energy transition and many modern technologies. However, the mining and refining process brings significant environmental and social challenges. This section explores the main problems and sustainable alternatives for cleaner mining.

Challenges in Mining Rare Earths and Critical Metals

Rare earth mining is particularly complex because these elements are not found in high concentrations and are usually mixed with other materials. Key challenges include:

- Highly polluting chemical separation processes

  - Extracted ores must undergo aggressive chemical treatments to separate usable metals, generating large volumes of toxic waste.

- Low element concentrations

  - Extracting rare earths requires removing large amounts of rock and soil for small quantities of the desired material.

- Dependence on a few countries

  - China dominates the sector, while other regions face economic challenges in developing mining capacity.

- Informal labor and exploitation

  - In countries like the Democratic Republic of Congo (DRC), cobalt is extracted under inhumane conditions, often involving child labor.

Environmental Impacts: Deforestation, Water Pollution, and Radioactive Waste

1. Deforestation and Soil Degradation

- Large-scale mining destroys natural ecosystems, especially tropical forests.

- Mine openings lead to soil erosion, making vegetation recovery difficult.

2. Water Contamination

- Chemical use in mineral separation contaminates rivers and aquifers.

- Elements such as thorium and uranium, often present in rare earth ores, can make water radioactive.

3. Toxic and Radioactive Waste

- Rare earth mining in China has generated massive ponds of highly polluting tailings.

- Poor waste management threatens biodiversity and local communities.

4. Health Impacts on Local Populations

- People living near mines often suffer from respiratory diseases and cancer due to exposure to heavy metals.

- Indigenous peoples and traditional communities are frequently displaced by mining expansion.

Sustainable Alternatives and Cleaner Extraction Technologies

1. Low-Impact Mining and More Efficient Processes

- Selective and precision mining: Advanced technology like geological sensors and AI allows more efficient, targeted extraction with reduced land disturbance.

- Dry processing: New techniques eliminate the need for large water volumes in refining, reducing contamination and waste.

- Use of biodegradable materials: Toxic chemicals are being replaced with gentler, biodegradable substances in separation processes.

2. Critical Mineral Recycling

- Reuse of electronic components: Large amounts of rare earths can be recovered from used batteries, EV motors, and discarded electronics.

- New recovery techniques: Companies are developing more efficient chemical extraction processes for recovering valuable metals from industrial and electronic waste.

- Circular supply chains: Encouraging recycling and reuse of critical metals can significantly reduce the need for new mining, extending resource life.

3. Deep-sea mining (with strict regulation)

- Controlled exploration of polymetallic nodules: The ocean floor holds vast critical metal reserves such as nickel, cobalt, and manganese. Companies are researching low-impact exploration methods.

- Low-impact robotic technologies: Autonomous underwater vehicles and mining robots are being developed to minimize disruption to marine ecosystems.

- Rigorous environmental monitoring: International regulations are being strengthened to ensure marine resource exploration is sustainable and transparent.

4. Biological Extraction and Nanotechnology (cutting-edge innovation!)

- Bioremediation in mining: Scientists are using bacteria and specialized microorganisms to dissolve minerals without harsh chemicals.

- Nanotechnology in mineral extraction: Nanomaterials are being used to separate ores more precisely, reducing losses and improving efficiency.

- Phytomining: Research is advancing on using hyperaccumulator plants to extract metals from soil—a process known as phytoextraction, which may be a sustainable alternative to traditional mining.

5. Transition to Renewable Energy in Mining

- Solar and wind power for mining operations: Companies are investing in renewable energy sources to reduce mining's carbon footprint.

- Energy storage for remote operations: Storage technologies allow mines to operate on clean power, even in off-grid locations.

- Lower CO_2 emissions: The use of electric vehicles and hydrogen-powered equipment is drastically reducing greenhouse gas emissions in the mining industry.

Chart 76: Comparison of Environmental Impacts: Traditional Mining vs. Sustainable Mining

Environmental Impact Comparison: Traditional vs. Sustainable Mining

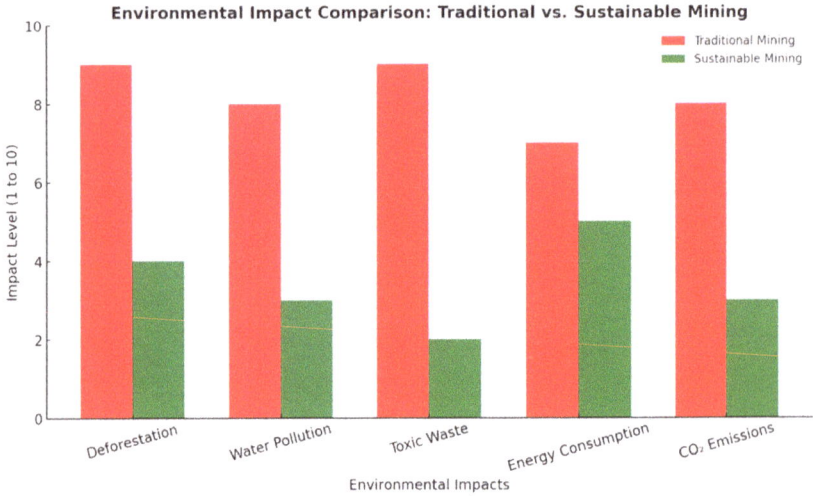

Source: Own elaboration based on the data presented in the Summary Table at the end of this chapter

Comparative chart showing the environmental impacts of traditional mining versus sustainable mining. It clearly illustrates how new technologies can significantly reduce environmental impacts.

Environmental Problem	Sustainable Solution
Deforestation	Selective mining and reforestation of affected areas
Water Pollution	Use of filters and bioremediation for water purification
Toxic Waste	Advanced waste management and recycling of critical minerals
Energy Consumption	Use of renewable energy sources in mining operations
CO_2 Emissions	Reduction of ore transport and energy efficiency in processing

Source: Own elaboration based on the data presented in the Summary Table at the end of this chapter

Difference Between Rare Earths and Critical Metals

Although the terms 'rare earths' and 'critical metals' are often used interchangeably, they have distinct definitions:

Rare Earth Elements: A group of 17 chemical elements in the Periodic Table, including lanthanum, neodymium, terbium, and dysprosium. They are widely used in electronic devices, electric vehicles, and wind turbines due to their unique magnetic and optical properties.

Critical Metals: A broader term that includes rare earth elements, lithium, cobalt, nickel, graphite, and other elements essential to modern technology. These metals are considered critical because of their high industrial importance and risk of scarcity or geopolitical dependence.

Critical metals are fundamental to the energy transition and technological innovation, requiring sustainable solutions for their extraction and refining.

Rare earth mining is undergoing a sustainable transformation driven by technological innovations and increased environmental responsibility. New practices and regulations are ensuring that the extraction of these essential minerals occurs in an increasingly clean and efficient manner. This allows the industry to continue supplying crucial materials for energy transition and technological innovation without compromising the environment and local communities.

The Future of Energy Transition

The energy transition is constantly evolving, driven by technological innovation and the search for sustainable solutions. The future of this sector will depend on scientific advances, new materials, and the strategic role of nuclear energy.

Technological Advancements and the Search for Sustainable Solutions

Technological innovation has been one of the fundamental pillars in accelerating energy transition. Key developments include:

- Environmentally responsible new extraction and refining processes: Advanced technologies are being developed to reduce the carbon footprint in the mining and refining of critical minerals.

- Improved recycling of strategic materials: New techniques make it possible to recover essential metals from electronic devices and used batteries, reducing the need for mineral extraction.

- Advances in batteries and energy storage: The development of solid-state batteries and new chemistries, such as sodium-ion batteries, may reduce dependence on critical minerals like lithium and cobalt.

- Growing use of green hydrogen: Hydrogen produced from renewable sources can replace fossil fuels in industrial sectors and energy storage.

- Artificial Intelligence and Big Data in energy optimization: The digitalization of energy systems improves efficiency in consumption and distribution, reducing waste.

Table 81: Rare Earth Applications in the Energy Transition

Element	Technological Use	Sector
Neodymium	High-power permanent magnets	Wind turbines, electric vehicles
Dysprosium	Thermal stability of magnets	Electric vehicles
Yttrium	Phosphors and superconductors	Solar panels, sensors
Terbium	Phosphors and magnetic devices	LED lighting, military tech

The Role of Hydrogen in Energy Transition

Hydrogen has emerged as an essential component in the search for sustainable energy solutions, offering alternatives to decarbonize various sectors of the economy. Its versatility allows applications ranging from electricity generation to use as a fuel in industrial processes and transportation.

Hydrogen production is classified based on the energy sources used and associated emissions:

- **Green Hydrogen**: Produced through water electrolysis using electricity from renewable sources such as solar and wind. This method generates no carbon emissions, making it a clean and sustainable option.

- **Blue Hydrogen**: Obtained from fossil fuels such as natural gas, with carbon capture and storage (CCS) technologies applied to reduce resulting CO_2 emissions.

- **Gray Hydrogen**: Also derived from fossil fuels, but without CCS, resulting in significant greenhouse gas emissions.

Hydrogen Applications in Energy Transition

Hydrogen has the potential to transform several sectors, helping to reduce carbon emissions:

- **Heavy Industry**: Sectors such as steel, cement, and chemicals can use hydrogen as a high-temperature heat source and raw material, replacing fossil fuels and reducing carbon footprints.

- **Transportation**: Hydrogen is promising for heavy-duty vehicles such as trucks, buses, trains, and ships, where direct electrification is challenging. Fuel cell vehicles offer long-range and fast refueling times.

- **Energy Storage**: Hydrogen can act as an energy carrier, converting excess electricity from renewables during low-demand periods into storable fuel that can later generate electricity or heat.

Challenges and Outlook

Despite its significant potential, widespread hydrogen adoption faces challenges:

- **Production Costs**: Green hydrogen production is still more expensive than traditional methods. Investment in R&D is essential to reduce costs and make hydrogen competitive.

- **Infrastructure**: A robust infrastructure for hydrogen production, storage, transportation, and distribution is necessary for effective integration into the energy system.

- **Energy Efficiency**: Hydrogen energy conversion processes involve losses, so improving their efficiency is crucial to maximize benefits.

Nevertheless, with advancing technologies and supportive policies, hydrogen is expected to play a central role in the transition to a cleaner and more sustainable energy matrix, aligning with global carbon reduction goals and climate action.

Hydrogen has stood out as a promising alternative to decarbonize the transport sector, offering energy efficiency and significant reductions in greenhouse gas emissions.

For example, vehicles powered by green hydrogen have shown GHG emission reductions of 87%, 85%, and 89% compared to the same vehicles fueled with diesel containing 7% biodiesel.

Furthermore, studies indicate that the efficiency of fuel cells powered directly by pure hydrogen is higher than those using hydrogen from hydrocarbon reforming.

These data highlight hydrogen's potential to improve energy efficiency and reduce emissions in the transportation sector.

Internal Combustion Engines (ICE):

CO_2 Emissions: Vehicles with ICEs using fossil fuels such as gasoline or diesel produce significant CO_2 emissions, contributing to atmospheric GHG levels.

Hydrogen Fuel Cell Vehicles (FCEVs):

CO_2 Emissions: When powered by green hydrogen (from renewable sources), FCEVs can reduce GHG emissions by up to 87% compared to diesel vehicles with 7% biodiesel.

Chart 77: Comparison of GHG Emissions Across Different Engine Types

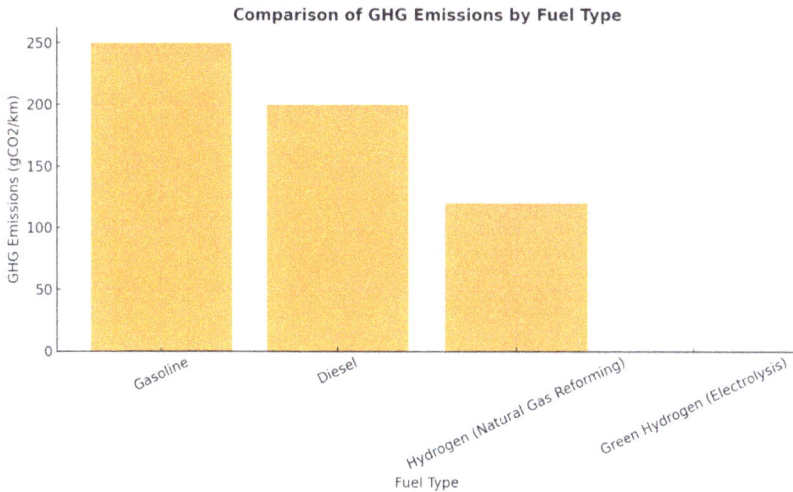

Comparison of GHG Emissions by Fuel Type

Source: Own elaboration based on the data presented in the Summary Table at the end of this chapter

Chart comparing greenhouse gas (GHG) emissions between combustion engines (gasoline and diesel) and hydrogen-powered engines (natural gas reformed hydrogen and green hydrogen via electrolysis).

The High Cost of Green Hydrogen and the Challenge of Renewable Intermittency

Green hydrogen, produced through water electrolysis using electricity from renewable sources, is often considered the most sustainable alternative among the different types of hydrogen. However, the chart shows that its production cost is significantly higher compared to gray and blue hydrogen. This high-cost stems largely from the reliance on renewable energy sources, which, although clean, present specific technical and economic challenges.

Comparison of Hydrogen Production Costs

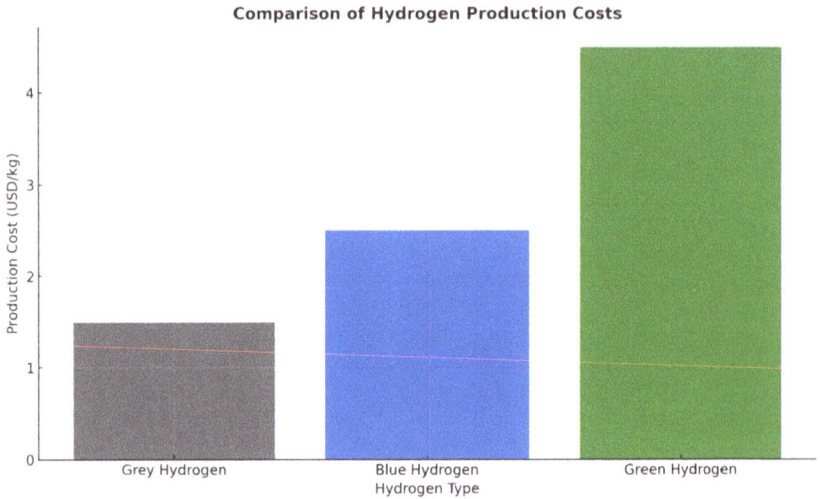

Source: Own elaboration based on the data presented in the Summary Table at the end of this chapter

Chart comparing the production costs of different types of hydrogen (gray, blue, and green) in dollars per kilogram.

Challenges of Green Hydrogen: Intermittency and Energy Efficiency

One of the main reasons for the high cost of green hydrogen is the intermittency of renewable energy sources. Solar and wind, for example, do not produce energy continuously, as they depend on variable weather conditions. This means that electrolysis may not operate at full capacity at all times, reducing operational efficiency and increasing costs. To overcome this limitation, solutions such as battery energy storage or integration with power grids are required to compensate for low-production periods—alternatives that further increase expenses.

Moreover, the infrastructure for green hydrogen production, storage, and transportation is still under development and lacks scale, which helps keep prices high. In contrast, gray hydrogen—produced from natural gas without carbon capture—remains the cheapest option because it uses established technologies and relies on a well-developed supply chain. Blue hydrogen, an intermediate option, has a higher cost than gray due to the carbon capture and storage (CCS) process but remains more competitive than green.

As technology advances and the cost of renewable energy continues to fall, green hydrogen is expected to become more affordable. However, this transition will require significant investment in infrastructure, research, and innovation to address the intermittency challenge and make green hydrogen economically viable on a large scale.

Energy Efficiency in Hydrogen Production: Amount of Electricity Required to Generate Hydrogen

One of the main challenges in hydrogen production—especially green hydrogen—is the amount of electricity needed to produce it. This factor directly impacts the process's economic viability and energy efficiency.

Water electrolysis, the method used to produce green hydrogen, has an energy efficiency ranging from 60% to 80% with the most advanced technologies. This means that for every 100 units of electricity consumed, only 60 to 80 are effectively converted into hydrogen—the rest is lost as heat and through other process inefficiencies.

In practice, producing 1 kg of hydrogen requires between 50 and 55 kWh of electricity. That 1 kg of hydrogen contains about 33.6 kWh of chemical energy, which highlights a considerable energy loss in the process. This efficiency can be compared to other energy storage methods like batteries, which have efficiencies greater than 90%.

How does this affect the viability of green hydrogen?

Electricity cost: Since electricity is the biggest cost in green hydrogen production, the process's economic feasibility depends heavily on the price of available renewable electricity. If electricity is expensive, hydrogen production becomes economically unviable.

Storage and conversion: After production, hydrogen must be stored and transported, processes that also consume energy and further reduce overall efficiency. Additionally, when hydrogen is reused to generate electricity (e.g., in fuel cells), another energy loss occurs, resulting in an overall efficiency of less than 40%—lower than battery-based systems.

Direct electricity use: In many applications, such as transportation, it may be more efficient to use electricity directly (via batteries) rather than converting it into hydrogen, especially if infrastructure supports it.

Chart 79: Efficiency in Green Hydrogen Production

Energy Losses in Green Hydrogen Production and Use

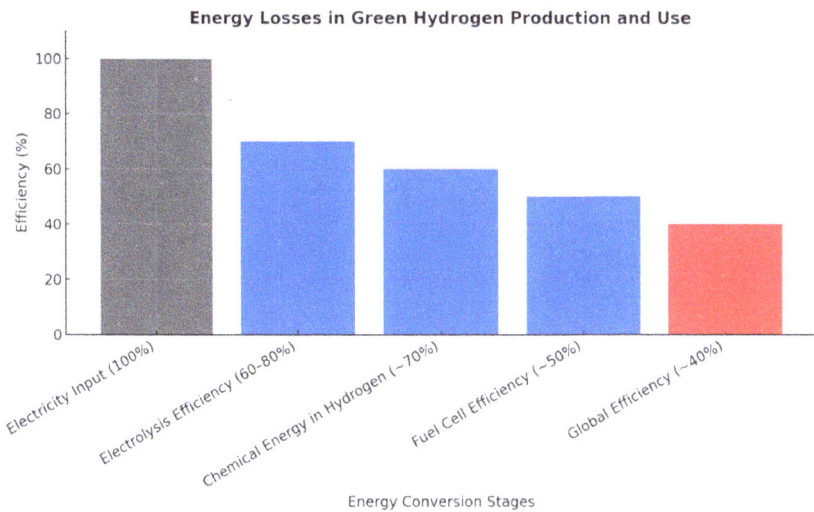

Source: Own elaboration based on the data presented in the Summary Table at the end of this chapter

Chart showing energy losses throughout the green hydrogen production and utilization process. It highlights how efficiency drops at each stage, from electrolysis to the final conversion of chemical energy into electricity.

Is There a Solution to Improve Efficiency?

Advances in high-temperature electrolysis and the use of surplus renewable electricity (when there is excess wind or solar generation) can significantly reduce costs and make the process more efficient. In addition, research into new catalysts and conversion technologies could improve energy efficiency rates in the future.

Green hydrogen production requires a considerable amount of electricity, with unavoidable losses throughout the process. This raises the question of when and where hydrogen is truly the best option compared to other energy solutions, such as battery storage. However, as the costs of renewable electricity fall and electrolysis technologies advance, the viability of hydrogen as a sustainable energy carrier may increase significantly.

The Possibility of New Materials Replacing Rare Earths in the Future

Research on Synthetic Substitutes for Rare Earths in Motors and Magnets

Rare earths are widely used in the production of high-performance electric motors and permanent magnets, essential for wind turbines, electric vehicles, and other high-tech devices. However, the dependence on these elements has driven research into alternative materials that can offer similar magnetic and electrical properties without the challenges of extraction and global supply.

- Ferrite magnets and advanced composites: Some research efforts aim to develop ferrite magnets doped with alternative elements, reducing the need for neodymium and dysprosium.

- Nanomaterials and synthetic alloys: Scientists are investigating hybrid compounds that mimic the behavior of rare earth magnets without relying on critical materials.

Alternative Materials That Reduce Dependence on Critical Minerals

Beyond magnets, other applications that use rare earths are being redesigned to depend less on these elements.

- Advanced metal alloys: Some studies explore iron-cobalt-based alloys that exhibit high coercivity and may replace neodymium-iron-boron (NdFeB) magnets.

- Transition elements as alternatives: Metals like manganese and cobalt are being studied as potential substitutes in various industrial applications.

- Advanced oxides and ceramics: Some sectors are investing in the development of ceramic materials that can perform functions similar to rare earths, particularly in catalysts and electronic systems.

Use of Superconductors to Eliminate the Need for Rare Metals in Certain Applications

A revolutionary approach to reducing rare earth dependence is the advancement of superconducting materials.

- Superconducting motors: Research into superconducting electric motors is gaining momentum, as these motors can operate with extremely high efficiency and without the need for permanent magnets.

- Room-temperature superconductivity: If room-temperature superconductors become viable, they could eliminate the need for several rare metals in electronic components and energy systems.

- Applications in energy generation and transmission: Superconducting cables and transmission systems could drastically reduce the use of rare earths in transformers and electrical equipment.

Chart 80: Comparison of Alternative Materials to Rare Earth Magnets

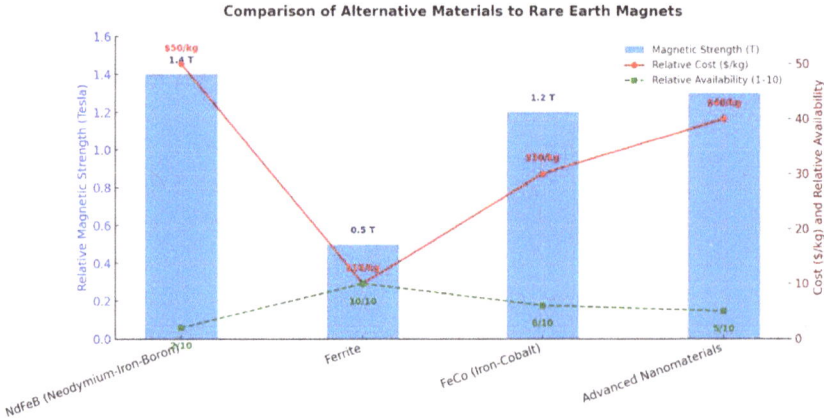

Chart comparing alternative materials to rare earth magnets, analyzing magnetic strength, cost, and availability. Neodymium-iron-boron (NdFeB) remains the strongest, but materials like iron-cobalt and advanced nanomaterials show potential as substitutes.

Table 82: Promising Research and Projects in Development

Institution/Company	Research/Project	Alternative Material	Development Stage
MIT & Toyota	Rare-Earth-Free Magnets	Iron-Nickel	Advanced Research
Hitachi	Electric Motors without Neodymium	Ferrite Alloys	Industrial Testing
Lawrence Berkeley Lab	Magnetic Nanomaterials	Composite Materials	Feasibility Studies
Cambridge University	Superconductors in Motors	Superconducting Wires	Experimental Development
GE Renewable Energy	Wind Turbines without Rare-Earth Magnets	Advanced Electromagnets	Field Testing

The Role of Uranium and Thorium in Low-Carbon Energy Generation

Nuclear energy plays a fundamental role in the energy transition as one of the few base-load electricity sources with low carbon emissions. The main fuels used in nuclear fission are uranium-235 and, to a lesser extent, thorium-232—both classified as strategic materials for global energy security.

- Uranium (U-235):

    - The primary nuclear fuel used in commercial reactors.

    - Requires enrichment to become usable in nuclear power plants.

    - Found in countries such as Kazakhstan, Canada, and Australia.

- Thorium (Th-232):

    - A promising alternative fuel, especially for molten salt reactors.

    - More abundant in the Earth's crust than uranium.

    - Countries like India, Brazil, and Norway have large reserves.

The advantage of both materials lies in their ability to provide reliable 24/7 energy without depending on weather variations that affect renewables, making them essential for the stability of the global power grid.

The Importance of Secure Supply Chains for Energy Stability

Energy security relies not only on nuclear technology but also on secure supply chains for uranium and other strategic materials. Uranium supply is concentrated in a few countries, making careful planning essential to avoid shortages.

Main uranium producers:

- Kazakhstan – the world's largest producer (around 40% of global output).

- Canada and Australia – key suppliers to the West.

- Niger and Namibia – critical sources for Europe.

Supply chain risks:

- Overreliance on a few countries can create vulnerabilities.

- Geopolitical instability may affect exports (e.g., sanctions on Russia, a major uranium processor).

- New mining and reprocessing projects can help reduce dependence on limited suppliers.

Beyond uranium and thorium, nuclear energy security also depends on critical technologies, such as the elements used in advanced fuels, cooling systems, and structural reactor materials.

Chart 81: Top Uranium Producers in the World

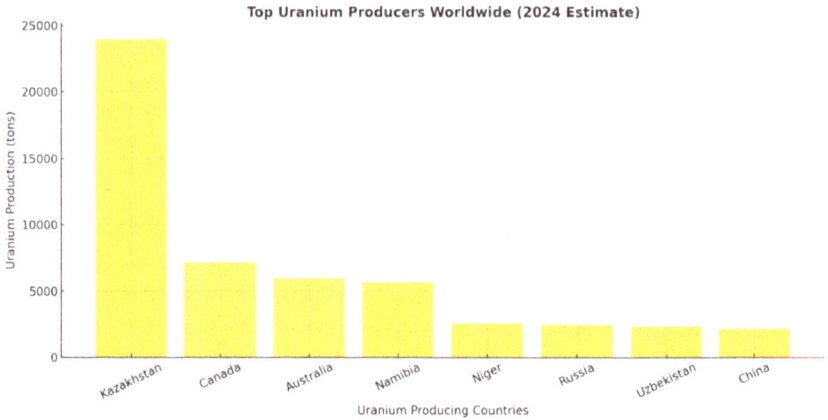

Top Uranium Producers Worldwide (2024 Estimate)

Source: Own elaboration based on the data presented in the Summary Table at the end of this chapter

How the Geopolitics of Critical Minerals Impacts Global Energy Security

The relationship between nuclear energy and energy security cannot be separated from the geopolitics of critical minerals. Many of the materials required for the safe operation of nuclear reactors are also strategically contested in the global market.

Key critical minerals in the nuclear industry:

- Zirconium – used in the cladding of fuel rods.

- Beryllium – used as a neutron moderator in some reactors.

- Lithium-6 – essential for fusion reactors and future applications.

The competition for these materials involves major global powers and can affect the stability of the energy market.

Dependence on supply chains dominated by a few countries—
such as China in the mining and refining of strategic materials—
can create new challenges for the nuclear sector and global
energy security.

Table 83: Critical Materials for the Nuclear Sector

Critical Material	Use in the Nuclear Industry	Main Suppliers	Supply Risks
Uranium	Fuel for nuclear reactors	Kazakhstan, Canada, Australia	Dependence on a few countries
Thorium	Alternative fuel for reactors	India, Brazil, Norway	Lack of infrastructure for commercial use
Zirconium	Fuel rod cladding	Australia, South Africa	Requires high nuclear-grade purity
Beryllium	Neutron moderator	USA, China, Kazakhstan	Limited and costly production
Lithium-6	Fusion reactors and military tech	China, USA, Russia	Restricted regulation and limited supply
Gallium	Advanced cooling systems	China, Germany, Russia	Production concentrated in China

Source: Own elaboration based on the data presented in the Summary
Table at the end of this chapter

GEO-POLITITICAL IMPACTS ON ENERGY

ENERGY RESOURCES

GEOPOLITICAL TENSIONS

UNSTABLE MARKETS

SUPPLY DISRUPTIONS

IMPACT ON ENERGY SECURITY

Conclusion of the Chapter – Energy Transition and the Role of Rare Earths

The energy transition toward a more sustainable and low-carbon future is closely tied to the availability of strategic materials, particularly rare earth elements and other critical minerals. The advancement of renewable technologies, batteries, electric vehicles, and even nuclear energy depends on the extraction, refining, and secure supply of these resources. However, this dependency brings economic, geopolitical, and environmental challenges that cannot be ignored.

The Centrality of Critical Minerals in the Energy Transition

The modern energy revolution is directly linked to elements such as neodymium, dysprosium, cobalt, and lithium, which are essential for producing high-performance magnets, long-life batteries, and wind turbines. Demand for these materials is growing exponentially, driven by global decarbonization targets and the electrification of key economic sectors. However, the geographic concentration of reserves and refining in a few countries—like China—creates a vulnerability scenario for the West and other dependent economies.

In Search of Alternatives: New Materials and Emerging Technologies

In response to supply risks and geopolitical tensions surrounding strategic minerals, investment in alternative materials is growing. Advanced research has already demonstrated the potential of new magnetic compounds and nanomaterials to reduce dependence on rare earths. Additionally, superconductivity and new industrial processes offer opportunities to mitigate the need for these raw materials in sectors like electric mobility and power generation.

At the same time, solutions such as strategic metal recycling and responsible mining can help ease environmental and geopolitical pressures, ensuring a more stable and sustainable supply of these resources.

Nuclear Energy and Strategic Minerals: A Pillar of Global Energy Security

The link between energy transition and global energy security cannot be separated from the role of uranium and thorium in electricity production. Nuclear energy remains a stable energy source with low carbon emissions and independence from weather variability—essential features for ensuring a resilient energy system. However, as with rare earths, the nuclear sector's supply chain is also sensitive to geopolitical factors, making it crucial to diversify uranium suppliers and invest in new technologies such as thorium reactors and nuclear fusion.

Geopolitics, Energy Security, and the Future of Energy Transition

Control over critical minerals and strategic fuels has become a central element of modern geopolitics, influencing trade disputes, national security policies, and international agreements. The balance between technological innovation, sustainable resource exploitation, and strategic partnerships will determine which nations lead the energy transition in the 21st century.

The pursuit of new mining models, resilient supply chains, and energy diversification is not merely an environmental issue. It is a geopolitical necessity to secure a stable and accessible energy future for all nations.

The energy transition is not just about replacing fossil fuels with renewables, it involves a new paradigm of exploration, dependency, and technological innovation. The future of energy is intrinsically linked to our ability to overcome the material, strategic, and geopolitical challenges that will shape the next era of human civilization.

Table 84: Sources Consulted in Chapter 9

Theme	Title	Author / Organization	Year	Link
Critical Minerals and Rare Earths	The Role of Critical Minerals in Clean Energy Transitions	IEA	2021	https://www.iea.org/reports/the-role-of-critical-minerals-in-clean-energy-transitions
Critical Minerals and Rare Earths	Critical Minerals Market Review 2023	IEA	2023	https://www.iea.org/reports/critical-minerals-market-review-2023
Critical Minerals and Rare Earths	Minerals for Climate Action	World Bank	2020	https://pubdocs.worldbank.org/en/961711588875536384/Minerals-for-Climate-Action-The-Mineral-Intensity-of-the-Clean-Energy-Transition.pdf
Critical Minerals and Rare Earths	Mineral Commodity Summaries 2023: Rare Earths	USGS	2023	https://pubs.usgs.gov/periodicals/mcs2023/mcs2023-rare-earths.pdf

Critical Minerals and Rare Earths	Study on the Critical Raw Materials for the EU 2023	European Commission	2023	https://data.europa.eu/doi/10.2873/725585
Critical Minerals and Rare Earths	2023 Critical Materials Assessment	U.S. DOE	2023	https://www.energy.gov/sites/default/files/2023-07/doe-critical-material-assessment_07312023.pdf
Nuclear Energy and Strategic Minerals	Uranium 2022: Resources, Production and Demand	OECD NEA / IAEA	2023	https://www.oecd-nea.org/jcms/pl_79960/uranium-2022-resources-production-and-demand
Nuclear Energy and Strategic Minerals	Thorium-Based Nuclear Energy: Options	IAEA	2023	https://www.iaea.org/publications/15215/near-term-and-promising-long-term-options-for-the-deployment-of-thorium-based-nuclear-energy
Nuclear Energy and Strategic Minerals	Thorium	World Nuclear Association	2024	https://world-nuclear.org/information-library/current-and-future-generation/thorium
Hydrogen	Global Hydroge	IEA	2023	https://www.iea.org/reports/global-hydrogen-review-2023

	n Review 2023			
Hydroge n	Green Hydroge n Cost Reductio n	IRENA	2020	https://www.irena.org/- /media/Files/IRENA/Agency/ Publication/2020/Dec/IRENA _Green_hydrogen_cost_2020. pdf
Hydroge n	The Future of Hydroge n	IEA	2019	https://www.iea.org/reports/t he-future-of-hydrogen
Emergin g Technolo gies	Substitut ion of critical raw material s in low- carbon technolo gies	Europe an Commi ssion JRC	2016	https://publications.jrc.ec.eu ropa.eu/repository/handle/JR C103284
Emergin g Technolo gies	Substitut ion strategie s for rare earths in wind turbines	Pavel et al.	2017	https://doi.org/10.1016/j.reso urpol.2017.04.010
Emergin g Technolo gies	Powering the green economy : magnets without	James McKenz ie / Physics World	2023	https://physicsworld.com/a/ powering-the-green- economy-the-quest-for- magnets-without-rare- earths/

	rare earths			
Environ mental & Social Mining Impacts	Assessin g social and environ mental impacts of critical minerals in Europe	Berthet et al.	2024	https://doi.org/10.1016/j.gloe nvcha.2024.102841
Environ mental & Social Mining Impacts	Mineral Resourc e Governa nce in the 21st Century	UNEP IRP	2020	https://www.resourcepanel.o rg/reports/mineral-resource-governance-21st-century
Environ mental & Social Mining Impacts	Myanma r's rare earth boom	Global Witness	2024	https://globalwitness.org/en/ campaigns/transition-minerals/fuelling-the-future-poisoning-the-present-myanmars-rare-earth-boom/
Recyclin g and Sustaina bility	Recyclin g of Critical Minerals	IEA	2023	https://www.iea.org/reports/r ecycling-of-critical-minerals
Recyclin g and Sustaina bility	Barriers to recycling rare earths in	Rizos et al. / CEPS	2024	https://www.ceps.eu/ceps-publications/understanding-the-barriers-to-recycling-

	energy transitio n			critical-raw-materials-for-the-energy-transition/
Geopoliti cs and Energy Security	Geopoliti cs of the Energy Transitio n: Critical Materials	IRENA	2023	https://www.irena.org/Public ations/2023/Jul/Geopolitics-of-the-Energy-Transition-Critical-Materials
Geopoliti cs and Energy Security	Energy Transitio n and Geopoliti cs: Critical Minerals	World Econo mic Forum	2024	https://www.weforum.org/pu blications/energy-transition-and-geopolitics-are-critical-minerals-the-new-oil/

Introduction – Final Chapter: The Role of Nuclear Energy in the Future of Humanity

Nuclear energy has always been at the center of debates about humanity's energy future. Throughout this book, we have explored its evolution, its challenges, and its promises for a more sustainable world. Now, as we approach the conclusion, it is essential to look ahead: what will be the role of nuclear in the coming decades? How will new technologies and scientific advancements shape this trajectory?

This final chapter brings together two essential fronts: the technological potential of nuclear energy for the future and its relevance to global energy sustainability. Here, we will address the following key topics:

- The new frontiers of nuclear technology, including Small Modular Reactors (SMRs), nuclear fusion, and artificial intelligence in the energy sector.

- Nuclear energy as a solution for the stability of the global energy matrix, ensuring energy security, low carbon emissions, and geopolitical independence.

- The balance between innovation, public acceptance, and government policy—factors that will determine whether the world moves toward a nuclear expansion or faces regulatory stagnation.

- The legacy of nuclear energy and its importance in the 21st century, showing how this energy source can be key to building a more sustainable, prosperous, and resilient civilization.

It is time to reflect on the overarching question that underpins this book: can the world truly guarantee a reliable energy future without nuclear energy?

Chapter 10 –
Conclusion:
The Role of Nuclear Energy in the Future of Humanity

What We Have Learned: The Evolution of Nuclear Energy

Throughout this book, we have followed a journey that mirrors humanity's own path in confronting the power of the atom — a path marked by scientific discoveries, political ambitions, technological breakthroughs, and profound ethical dilemmas.

From the earliest laboratory experiments in the early 20th century to today's cutting-edge next-generation reactors, nuclear energy has evolved from a scientific curiosity into a transformative force — one capable of both destruction and illumination. We have learned how the fission of the atomic nucleus — discovered by Hahn, Strassmann, Meitner, and Frisch — sparked not only the nuclear arms era but also the beginning of a new way to generate electricity on a large scale.

We witnessed the birth of the nuclear age under the shadow of World War II and the Manhattan Project, but we also saw how, in the decades that followed, the world sought to reverse this military legacy for peaceful purposes — with initiatives such as the "Atoms for Peace" program, the construction of the first nuclear power plants, and the creation of the IAEA.

We observed the rise and fall of the nuclear sector: from its global expansion in the 1960s and 70s, through the impact of the accidents at Three Mile Island and Chernobyl, to the post-Fukushima recovery of trust and the renewed interest driven by the urgency of the climate crisis.

Today, in the face of energy transition challenges, geopolitical instability, and the urgent need to deeply decarbonize the global economy, nuclear power is reemerging — not as a replacement but as an essential complement to renewable sources. It is not a choice between the sun, the wind, or the atom — it is an intelligent integration of all available options grounded in science and common sense.

But perhaps the most important lesson we have learned is that nuclear energy evolves — and will continue to evolve. The history of fission is already long, but the future may lie in fusion. What was once synonymous with centralization and massive infrastructure now points toward modularity, intrinsic safety, and flexibility.

Nuclear energy is not a relic of the past — it is a technology in constant reinvention, one that challenges myths and transcends ideologies. By understanding its trajectory, we also understand the enormous potential it holds to help build a safer, more stable, and more sustainable energy future for all of humanity.

New Technologies and the Future of the Nuclear Sector

Nuclear energy is undergoing a quiet yet highly promising revolution. Far from the traditional image of massive reactors built in the 20th century, new technologies are reshaping the sector with a focus on safety, efficiency, flexibility, and environmental sustainability. This new era is defined by three fundamental vectors: Small Modular Reactors (SMRs), real advances in nuclear fusion, and the growing use of artificial intelligence, robotics, and advanced materials.

Small Modular Reactors (SMRs) and Energy Decentralization

SMRs represent one of the most promising innovations in the contemporary nuclear industry. As the name suggests, they are smaller-scale nuclear reactors designed to produce between 10 and 300 megawatts of electricity (MWe), compared to over 1000 MWe in conventional nuclear power plants.

Their greatest advantage lies in modularity and flexibility: they can be manufactured in series within controlled industrial environments and then transported to the installation site, drastically reducing costs, construction time, and the risks associated with civil works.

These reactors allow for:

- Decentralizing energy production, delivering electricity to remote regions or islands.

- Integration into hybrid grids alongside renewable sources like solar and wind.

- Replacing coal-fired thermal power plants with lower environmental impact.

- Serving specific industrial applications such as desalination, hydrogen production, or process heat for industrial use.

Countries like Canada, the United States, the United Kingdom, and France already have advanced SMR programs. Companies such as NuScale Power (USA), Rolls-Royce SMR (UK), and Terrestrial Energy (Canada) are leading the race. The first commercial SMRs are expected to begin operation within this decade.

The Progress of Nuclear Fusion: Challenges and Real Prospects

Nuclear fusion is the "Holy Grail" of energy: clean, safe, abundant, and virtually inexhaustible. Instead of splitting heavy atoms like uranium, fusion consists of joining light nuclei — usually hydrogen isotopes such as deuterium and tritium — releasing a tremendous amount of energy, just as it occurs in the core of the Sun.

Despite decades of research and delayed promises, recent years have brought concrete advances:

- In 2022, scientists at the National Ignition Facility (USA) achieved ignition for the first time, generating more energy than was consumed by the laser.

- In 2024, China announced a new record with its EAST reactor ("Artificial Sun"), sustaining plasma at 158 million degrees Celsius for over 1,000 seconds.

- International projects such as ITER in France continue to make progress, with operations expected to begin between 2025 and 2030.

EAST
(Experimental Advanced Superconducting Tokamak)

- Location: China
- Purpose: Research
- Temperature: 158 million °C
- Plasma duration: >1,000 s
- Goal: Plasma sustainment and control

ITER
(International Thermonuclear Experimental Reactor)

- Location: France (International project)
- Purpose: Commercial demonstration
- Forecast: First plasma in 2025
- Goal: Produce 10x more energy than it consumes

Comparative illustration between the EAST and ITER projects:

Source: Own elaboration based on the data presented in the Summary Table at the end of this section.

- EAST, from China, is an advanced research project focused on plasma maintenance and high-temperature control.

- ITER, in France, is the largest international effort to demonstrate the viability of fusion as a commercial energy source.

The main barrier to fusion is no longer scientific feasibility — already demonstrated — but rather technical and commercial viability. Keeping plasma stable, containing neutron radiation, and developing materials that can withstand extreme environments remain complex technical challenges.

Nonetheless, private startups such as Commonwealth Fusion Systems, TAE Technologies, and the UK-based Tokamak Energy promise commercially viable fusion reactors by 2040.

Chart 82: Nuclear Fusion Timeline – Key Historical Milestones

Timeline of Nuclear Fusion: Historical Milestones

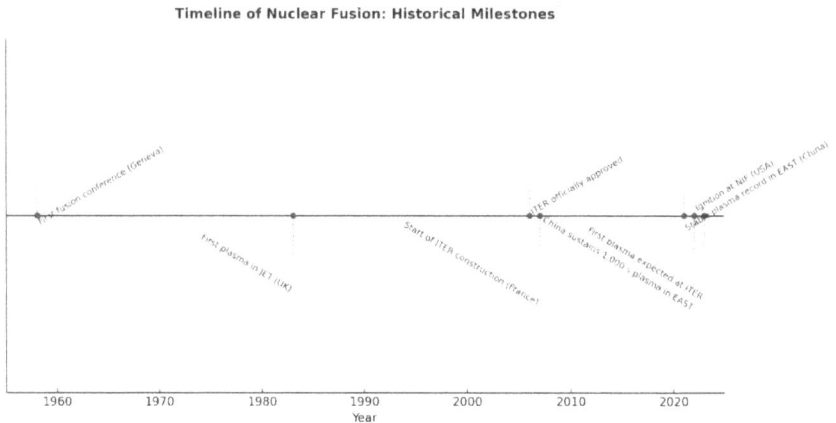

Source: Own elaboration based on the data presented in the Summary
Table at the end of this section

AI, Robotics, and New Materials in the Next Generation of Reactors

The new nuclear era will also be profoundly shaped by digital technologies and breakthroughs in materials science:

- Artificial Intelligence (AI) is already being used for nuclear simulations, failure prediction, performance optimization, and real-time predictive maintenance.

- Robots and drones perform internal inspections in radioactive environments with greater safety and precision, reducing human exposure.

- Advanced materials, such as corrosion-resistant metal alloys and high-temperature ceramics, are enabling the development of fourth-generation reactors that are safer and more durable.

Moreover, the development of self-healing materials, smart coatings, and embedded sensors in structural components is paving the way for self-managing reactors with automated maintenance and adaptive responses to anomalies.

Technology Flowchart: AI, Robotics and Materials in Next-Generation Reactors

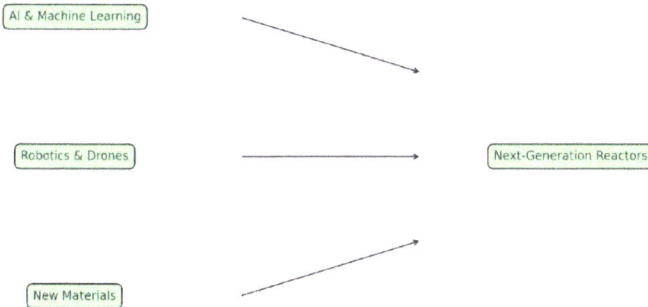

Source: Own elaboration based on the data presented in the Summary Table at the end of this section

The technological flowchart illustrates how AI, robotics, and new materials converge to drive next-generation nuclear reactors:

- *AI supports control, simulation, and predictive maintenance.*

- *Robotics and drones ensure safe and automated inspections.*

- *New materials withstand extreme environments with greater durability and safety.*

A Future in the Making

By combining these three fronts — SMRs, fusion, and emerging technologies — nuclear energy could play a vital role in the energy transition of the 21st century. Its ability to provide firm, clean, and reliable power will be essential to complement renewable sources and ensure energy security, deep decarbonization, and strategic autonomy for nations.

The Chinese Case: Nuclear Leadership in the New Energy Era

China has emerged as a global leader in the nuclear energy landscape through cutting-edge initiatives. In March 2025, the country announced the successful operation of its first thorium-fueled molten salt reactor located in the Gobi Desert. This type of reactor uses thorium instead of uranium and offers several significant advantages, such as greater intrinsic safety, lower risk of core meltdown, and reduced production of long-lived radioactive waste.

Additionally, China is advancing in the field of nuclear fusion, attempting to replicate the processes of the Sun here on Earth. Its EAST project — Experimental Advanced Superconducting Tokamak, also known as the "artificial sun" — set a new world record by maintaining stable plasma for 1,066 seconds — an impressive milestone and a crucial step toward the future viability of fusion as a clean and virtually inexhaustible energy source.

These developments position China at the technological forefront of the nuclear sector, both in fundamental research and practical implementation, with potentially transformative geopolitical and energy impacts in the decades to come.

Nuclear Energy and Global Sustainability

In this section, we will explore the role of nuclear energy as an ally to renewables, its contribution to energy security, and its potential as a concrete solution to the climate crisis. We will also highlight the importance of stable public policies and long-term strategies to ensure the sustainable and safe development of the sector.

The climate emergency and the energy transition have placed the world before a paradox: how to decarbonize rapidly without compromising energy supply and economic stability? In this context, nuclear energy is reemerging as one of the few available technological solutions truly capable of generating large-scale electricity without carbon emissions and with high reliability.

Coexistence with Renewables and Energy Security

Renewable sources, such as solar and wind, have grown impressively over the past two decades. However, their natural intermittency — the sun doesn't shine at night, and the wind doesn't always blow — demands complementary solutions to ensure a continuous and stable energy supply.

This is where nuclear energy stands out:

- Provides firm (baseload) energy 24 hours a day, regardless of weather conditions.

- Can be integrated into hybrid systems alongside renewables, stabilizing the power grid.

- Reduces reliance on fossil fuel backup sources like natural gas or coal.

- Helps prevent blackouts and supply crises, especially during peak demand or prolonged droughts (which impact hydroelectric generation).

Examples such as France — which maintains one of the world's cleanest and most stable electricity grids thanks to its nuclear matrix — show that coexistence with renewables is possible, enabling a balanced and resilient transition.

The Importance of Long-Term Policies

Unlike other energy technologies, nuclear energy requires long-term strategic planning at both technical and political levels. The construction, licensing, and operation of a nuclear power plant involve decades of work, requiring:

- Regulatory and legal stability.

- Consistent public and private investment.

- Specialized technical training and preservation of know-how.

- Public acceptance and the fight against misinformation.

The absence of clear policies has led many countries to abandon or postpone nuclear projects — only to later face energy crises and return to fossil fuels.

Conversely, countries like Finland, South Korea, Canada, and China have shown that with well-structured policies, it is possible to expand nuclear energy safely, efficiently, and transparently, contributing significantly to climate goals.

Nuclear Energy as a Viable Climate Solution

According to the Intergovernmental Panel on Climate Change (IPCC), nuclear energy is essential in nearly every plausible scenario to limit global warming to 1.5°C. It is one of the few low-emission sources with sufficient technological maturity to replace fossil fuels at scale.

Its advantages are clear:

- Virtually zero CO_2 emissions during operation.

- A carbon footprint comparable to wind energy — and lower than solar — over its full lifecycle.

- High energy output with minimal land use, unlike some renewable sources.

- Potential to contribute to clean hydrogen production, industrial process heat, and desalination.

Even with challenges related to waste and safety (already addressed in previous chapters), nuclear energy remains an indispensable tool in the global climate arsenal.

Table 85: Expected Nuclear Contribution in Carbon Neutrality Scenarios (2050)

Scenario	Source	Expected Nuclear Capacity (GW)	Share of Global Electricity Generation
Net Zero by 2050	IEA	812 GW	~18%
Sustainable Development Scenario	IEA	700 GW	~15%
MIT Nuclear Study	MIT	1000 GW	~20%
IPCC SSP2-1.9 Scenario	IPCC	900 GW	~16–20%

Source: Own elaboration based on the data presented in the Summary Table at the end of this chapter

Table 86: Emerging Technologies with Nuclear Potential

Technology	Current Status	Potential Impact
Small Modular Reactors (SMRs)	Testing and licensing stage	Decentralization and enhanced safety
Nuclear Fusion (ITER, DEMO)	Prototyping and testing	Unlimited and clean long-term energy
Advanced Fuel Recycling	Pilots in France and Japan	Waste reduction and efficiency increase

Source: Own elaboration based on the data presented in the Summary Table at the end of this chapter

Final Conclusion: Nuclear Energy as a Pillar of the 21st Century

For centuries, humanity has dreamed of mastering energy. From fire to the steam engine, from coal to oil, from electricity to the atom, we have walked a fascinating path filled with discoveries, mistakes, and lessons. Today, on the brink of climate and energy collapse, we are faced with a decisive choice: to continue relying on insufficient solutions or to boldly embrace the tools that can truly ensure a sustainable future.

Nuclear energy — long misunderstood and unjustly dismissed — is one of those tools. It is not a panacea. It is not risk-free. But as we have shown throughout this book, it is one of the most powerful, clean, and effective technologies ever conceived by humankind. When well-planned, well-regulated, and well-communicated, it can coexist with renewables, stabilize power grids, drastically cut emissions, and guarantee long-term energy sovereignty.

But for that promise to be fulfilled, myths, prejudices, and political resistance must be overcome. Disinformation, fear, and ideological manipulation have proven to be obstacles as dangerous as any technical flaw. Many governments have turned their backs on science, surrendering to simplistic narratives and jeopardizing the energy future of the next generations.

The time for change has come.

There is no serious energy transition without commitment. No carbon neutrality without firm, clean electricity. No sustainable civilization without the courage to face the facts.

The new generation of reactors, the advances in fusion, the integration with digital technologies, and the use of innovative materials are proof that nuclear is not the past — it is the future. A future that demands innovation with safety, progress with responsibility, and science with ethics. And above all, it demands political leadership.

This is where you, the reader, come in.

This book was not written merely to inform. It was written to call upon you. If you've reached this point, you now know that nuclear energy is not just one more option among many. It is a strategic choice for humanity's survival on a finite planet. And as such, it must be discussed seriously, promoted honestly, and implemented competently.

It is time to make it clear to political leaders — in parliaments and in international forums — that society demands real solutions, not empty narratives. That the right to a habitable planet also entails the duty to support science and reason.

Nuclear energy is not the enemy. The enemy is ignorance. It is inaction. It is delay disguised as virtue.

The 21st century will be shaped by the choices we make now. And among those choices is the place we give to nuclear energy.

Let every reader become a voice. Let every voice become action. And from that action, may a future arise in which the light that powers our homes, our factories, and our hospitals come from a source that is clean, safe, powerful, and, above all, honest.

Because the future is not predicted — it is built.

And the time to build it is now.

Table 87: Sources Consulted in Chapter 10

Topic	Source	Link
Reator de Tório (China)	O Cafezinho	https://www.ocafezinho.com/2025/03/23/china-anuncia-primeira-usina-nuclear-de-sal-fundido-com-torio-do-mundo/
Reator de Tório (China)	digitalagro.com.br	https://digitalagro.com.br
Fusão Nuclear (EAST – China)	ICL Notícias	https://iclnoticias.com.br/china-sol-artificial-ocidente-para-tras/
Fusão Nuclear (EUA – NIF)	Wikipedia	https://pt.wikipedia.org/wiki/National_Ignition_Facility
Projetos SMRs	IAEA	https://www.iaea.org/topics/small-modular-reactors
Fusão – ITER	ITER Organization	https://www.iter.org
IA e Robótica em Reatores	World Nuclear Association	https://www.world-nuclear.org
Coexistência Nuclear-Renováveis	World Nuclear Association	https://www.world-nuclear.org
Segurança Energética e Baseload	IEA – International Energy Agency	https://www.iea.org

Políticas de Longo Prazo	OECD-NEA	https://www.oecd-nea.org
Exemplos Políticas Nucleares	IAEA	https://www.iaea.org
Nuclear e Clima	IPCC	https://www.ipcc.ch
Pegada de Carbono	World Nuclear Association	https://www.world-nuclear.org/information-library/current-and-future-generation/nuclear-power-and-the-environment.aspx
Importância da Energia Nuclear no Século XXI	IPCC	https://www.ipcc.ch
Transição Energética e Neutralidade Carbónica	IEA	https://www.iea.org

www.ingramcontent.com/pod-product-compliance
Lightning Source LLC
Chambersburg PA
CBHW041206220326
41597CB00030BA/5061